DIAGNOSIS:
PHILOSOPHICAL AND MEDICAL PERSPECTIVES

Episteme

A SERIES IN THE FOUNDATIONAL,
METHODOLOGICAL, PHILOSOPHICAL, PSYCHOLOGICAL, SOCIOLOGICAL, AND
POLITICAL ASPECTS OF THE SCIENCES, PURE AND APPLIED

VOLUME 15

The titles published in this series are listed at the end of this volume.

DIAGNOSIS:
PHILOSOPHICAL AND MEDICAL PERSPECTIVES

by

NATHANIEL LAOR

Tel-Aviv-Brüll Community Mental Health Center,
Tel-Aviv University
and
Child Study Center, Yale University

and

JOSEPH AGASSI

Department of Philosophy, Tel-Aviv University
and
Department of Philosophy, York University,
Toronto

KLUWER ACADEMIC PUBLISHERS
DORDRECHT / BOSTON / LONDON

Library of Congress Cataloging in Publication Data

Laor, Nathaniel.
   Diagnosis : philosophical and medical perspectives / by Nathaniel
Laor and Joseph Agassi.
      p.   cm. -- (Episteme ; v. 15)
   Includes bibliographical references (p.   ) and indexes.
   ISBN 0-7923-0845-X (acid-free paper)
   1. Diagnosis. 2. Nosology. 3. Diagnosis--Philosophy.
I. Agassi, Joseph. II. Title. III. Series: Episteme (Dordrecht,
Netherlands) ; v. 15.
RC71.L265 1990
616.07'5'01--dc20   90-39957

ISBN 0-7923-0845-X

Published by Kluwer Academic Publishers,
P.O. Box 17, 3300 AA Dordrecht, The Netherlands.

Kluwer Academic Publishers incorporates
the publishing programmes of
D. Reidel, Martinus Nijhoff, Dr W. Junk and MTP Press.

Sold and distributed in the U.S.A. and Canada
by Kluwer Academic Publishers,
101 Philip Drive, Norwell, MA 02061, U.S.A.

In all other countries, sold and distributed
by Kluwer Academic Publishers Group,
P.O. Box 322, 3300 AH Dordrecht, The Netherlands.

Printed on acid-free paper

Printed in the Netherlands

TABLE OF CONTENTS

ABSTRACT vii

ACKNOWLEDGEMENT ix

PREFACE xi

Chapter 1: INTRODUCTION 1

Chapter 2: THE PROBLEM SITUATION 25

Chapter 3: THE COMPUTER REVOLUTION 41

Chapter 4: ETHICS OF DIAGNOSTIC SYSTEMS . . 67

Chapter 5: SYSTEMS AND MEDICINE 91

Chapter 6: DIAGNOSTIC THEORY 115

Chapter 7: DIAGNOSTIC PRACTICE 137

Chapter 8: SOME INTERFACES 155

Chapter 9: THE HUMAN FACTOR 183

Chapter 10: CONCLUSION: 209

SELECTED BIBLIOGRAPHY FOR FURTHER
      READING 235

INDEX OF NAMES 249

INDEX OF SUBJECTS 251

ABSTRACT

The place of medical diagnosis in the chart of medical care is becoming increasingly central and problematic; this invites a comprehensive view of diagnosis in its context, and this requires multi-dimensional theoretical analyses of the diagnostic system and its interfaces. Here is a first venture on this path which will hopefully appeal to those concerned with the general aspects involved here, from the philosophy of science, of technology and of medicine, as well as to those concerned with the special aspects involved here, from medicine and allied fields, systems theory, computer technology, and public administration.

The goal of this book is threefold: (1) scientific: to present a comprehensive framework for medical diagnosis conducive to the study of specific diagnoses; (2) technological: to identify limits inherent to the diagnostic process in general and to computer-assisted diagnosis in particular, as well as to advocate and help design a nation-wide comprehensive computer-assisted medical diagnostic service; and (3) ethical: to mobilize public controls for the protection of physicians and patients against the misuse of diagnosis, especially when computer-assisted. To these ends patient informed consent should be sought throughout the diagnostic process so as to invite patients to participate in their own curative processes, thus restoring the traditional association of medicine and education; the computer should be recruited to enhance this restoration.

The discussions in this book forgo technical jargon with no loss of content while negotiating in context diverse problems associated with a wide range of topics. The book is readily accessible to many audiences and may serve as a primary resource in graduate and post-graduate studies of beginners and adepts in the fields of medicine, of systems theory and of philosophy--as each applies to the young and developing field of diagnostics.

ACKNOWLEDGEMENT

Nathaniel Laor worked on this book while training in systems analysis in the Israeli Defense Force, winter 1978, and completed the work during 1987-1988, while serving on the faculty at the Department of Psychiatry, Yale School of Medicine and Yale Child Study Center. Generous comments and support were offered by Boris M. Astrachan, of Yale School of Medicine, Eran Dolev, the former Surgeon General of the Israeli Defense Force, currently at Sackler School of Medicine, Tel-Aviv University, and Paul Ererra, of Yale School of Medicine.

Joseph Agassi worked on this book while a senior fellow of the Alexander von Humboldt-Stiftung--one year (1977-8) in the Zentrum für interdisziplinäre Forschung, Universität Bielefeld, and one year (1987) in the Department of Economics, Johann Wolfgang Goethe Universtät Frankfurt.

Earlier drafts of this book were read critically by Mario Bunge of McGill University, I.C. Jarvie of York University, Toronto, Michael Rosenbluth of the University of Toronto School of Medicine, Charles M. Sawyer of Prime Computer and Harvard University Extension School, Jacob Steif, Deputy Director for Research and Development of Hadassah University Hospital, and John Strauss of Yale School of Medicine.

In the endless process of correction, Etta Burke, Kitty Moore and Dorine Perach supplied much patient assistance.

Our gratitude to them all.

PREFACE

1. GENERAL

The term "diagnostics" refers to the general theory of diagnosis, not to the study of specific diagnoses but to their general framework. It borrows from different sciences and from different philosophies. Traditionally, the general framework of diagnostics was not distinguished from the framework of medicine. It was not taught in special courses in any systematic way; it was not accorded special attention: students absorbed it intuitively.

There is almost no comprehensive study of diagnostics. The instruction in diagnosis provided in medical schools is exclusively specific. Clinical instruction includes (in addition to vital background information, such as anatomy and physiology) specific instruction in nosology, the theory and classification of diseases, and this includes information on diagnoses and prognoses of diverse diseases. What is the cause of the neglect of diagnostics, and of its integrated teaching? The main cause may be the prevalence of the view of diagnostics as part-and-parcel of nosology. In this book nosology is taken as a given, autonomous field of study, which invites almost no comments; we shall freely borrow from it a few important general theses and a few examples.

We attempt to integrate here three studies:
1) of the way nosology is used in the diagnostic process;
2) of the diagnostic process as a branch of applied ethics;
3) of the diagnostic process as a branch of social science and social technology.

Computer technology is used as a central example for each of these three studies: we consider the introduction of the computer into the field of medicine in general, and into the diagnostic encounter in particular, both a great step forward and a great risk--for patients, for physicians, and for the public at large.

The present neglect of the study in diagnostics is reinforced by the popularity of the claim that theory (nosology and diagnostics included) should emerge from observed facts. This claim is the myth of science as inductive; it will be repeatedly discussed here. One of its major ill-effects is in its discouragement of the explicit presentation of views in preference for embedding them in case-studies; readers of case-studies are expected to absorb intuitively the theories they embed. Since the explicit and systematic presentation of a theory facilitates its critical examination, the myth of induction makes it hard to examine new theories critically.

The need for a systematic critical thinking about diagnostics is growing for a number of reasons. *First*, this century witnessed profound reform in the philosophy of science and technology. *Second*, the immense growth of the diverse highly specialized sciences which contribute to medical practice in general, and to diagnostics in particular, calls for a check-list of all that enters the field and for the critical examination of whether each item is updated at reasonably frequent intervals. *Third*, outside the natural sciences and the life sciences new scientific theories were developed, whose importance for medicine is increasing. *Fourth*, the introduction of computer technology to diagnostics has effected a quiet revolution that calls for a thorough assessment of its benefits and dangers, and for public control of the quality of practices which are deemed considerably risky. These four items invite detailed and explicit articulations as well as systematic and critical examinations.

The rest of this *Preface* should help the reader develop from the start a picture of the situation as it will be presented here.

2. SPECIFIC

Philosophy

Reformers of medical practice can hardly avoid paying attention to philosophical innovations, as some of these render obsolete two scientific philosophies embraced by medical tradition. One is mechanism or mechanist materialism, the idea that wholes consist of sums of their parts and therefore can be described in terms of nothing but parts that can be clearly delineated. This has been replaced with systemism or systems theory or the systems approach or systems-theory materialism (Mario Bunge). The other is inductivism or extreme empiricism, the idea that theories emerge from less abstract theories and ultimately from observations of facts. This has been replaced with hypothetico-deductivism or with deductive nomological empiricism or attenuated empiricism (William Whewell). The idea that scientific theories are true or at least highly probable has been replaced by the idea that science is a series of theories which are approximations to the truth (Albert Einstein). Scientific certitude has been replaced with scientific openness to critical examination (Karl Popper).

Inventory of diagnostic tools

The current use of an ever increasing wealth and complication of scientific and technological background knowledge in medicine makes it increasingly important to draw a check-list or an inventory of the scientific tools available in the battle against disease. This is particularly true regarding common diagnostic tools which have obvious functions. The function of the microscope, the stethoscope, or the ophthalmoscope, seems unproblematic, as each broadens the diagnostician's field of vision. Yet, notoriously, an untrained clinician using these instruments easily misperceives whatever the instruments are supposed to help perceive. The same holds for modern pattern-recognition computer-assisted tools, such as the celebrated computerized axial tomographic scanner (CAT scan). Expertise is needed in order to read images properly.

Inventory of relevant new sciences

New scientific specialties are now offering medicine new and powerful tools. These include decision theory, storage and queuing theory, linear programming, opportunity-cost theory, cost-effectiveness theory, systems analysis and public administration. These have significant applications to medicine in general and to diagnostics in particular, even though they have become increasingly problem-ridden. It is not easy to see

how queuing theory effects a single diagnostic encounter in a
private clinic; yet it does, and at times with a vengeance.
Physicians often employ intuitive considerations though they
have access to much better and significantly more precise
modes of reasoning. At times intuitive considerations impede
the diagnostic process, especially when they have severe adverse
impact on the administrative and economic environment in
which they take place.

The computer in the service of medicine

The arrival of the computer into the service of diagnosis
presents yet another reason for the urgent need for an explicit,
systematic study of diagnostics. In the strict sense, the idea of
a computerized diagnostic encounter was an attempt to replace
diagnosticians by robots. This was seriously undertaken as
soon as computers became available, but was given up a few
years ago. It was replaced by attempts to offer the general
market computer software to serve in computer-assisted
diagnostic services, better known as diagnostic expert systems.
These attempts have thus far achieved only partial success. It
was spectacular, though. The list of successes is relatively
short, and the list of spectacular ones is naturally shorter;
moreover, a spectacular success may easily be discarded a short
time after its introduction as not having lived up to its
promise. One example of spectacular success of using
computers in diagnosis is this: digitalis is a traditional toxic
medication for heart condition. Patients who use it have to
pass a test for toxicity regularly. A certain computer program
greatly improved the ability of heart specialists to perform that
test accurately. Likewise, physicians consulting a computer
program for the diagnosis of infectious diseases were reported to
have almost matched expert levels. We find more challenging
the less spectacular computerized service for heart patients,
providing them by telephone rough diagnoses and suggestions.

3. AN OVERVIEW

In all practical processes, diagnosis included, the guiding
concern is pragmatic. In a study of a pragmatic concern,
pragmatic needs invite theoretical consideration, though these

may be limited to specific uses. In the study of diagnosis the use of theory was generally unproblematic; it is no longer so, since nowadays relevant theoretical material comes from many fields. This includes

(1) theories scattered within the medical field, collated and subjected to thorough critical examination and assessment;
(2) a general philosophical and psychological framework for decision processes (as applied to medicine);
(3) the general abstract apparatus of decision theory, pattern-recognition theory (including information theory and data processing theory), systems analysis (including operational research and, most significantly, the parts of economics which concern opportunity cost and cost effectiveness);
(4) ethical theory concerning the private and the public domains.

Ethical theory has seldom been systematically applied to concrete problems. The advent of radically new technologies has forced the medical profession to seek systematic guidance. In response, some philosophers have inaugurated a new field of moral guidance, biomedical ethics, which has acquired some of the glitter of the medical hardware with which it is associated. We will ignore it whenever possible: there are enough moral problems on our agenda. Rather, we shall take medical ethics as inherent to diagnosis. We shall address paternalism, the readiness physicians exhibit to take responsibility and make decisions which by right and by duty belong to patients. We will not philosophize about paternalism nor moralize about it, but offer what we hope is potent medicine against some of its harsh manifestations. There is a vast literature for and against paternalism, and much of it concerns medicine. We shall bypass most of this literature, since usually (we shall mention the exceptions) it suffers from a serious defect: it almost entirely overlooks paternalism in diagnosis. Moreover, if it does not endorse paternalism it endorses the classical alternative to it, extreme individualist ethics. Both alternatives are untenable: the defects of each reinforce the other. They must both be replaced by a more systemic ethical approach: responsibility is always individual, yet the setting of even the most private clinical diagnostic encounter is public.

Our presentation is largely methodological, and exhibits two methodological biases in addition to the ethical one just mentioned. *First,* we have a new view of scientific technology and its domain of applicability which eschews the concept of scientific provability or probability in favor of responsible conduct as judged by legal or customary standards, which should be subject to improvement through the usual methods of democratic reform. *Second,* we avoid the two ancient approaches, based on mechanism and holism, the two classical views on the nature of things, the views of those who proceed from the detail to the general picture and those who go the opposite way; we replace them with a new approach based on compromise and common sense. Quite generally, as certainty is no longer required, different methods may be tentatively used at different times in the hope that this will improve output. When competing methods are based on mechanism or on holism, often the practical situation prescribes choice between them. Most surprisingly, yet in accord with common sense, some aspects are at times sufficiently isolated and are amenable to treatment in accord with the mechanist view; other aspects have considerable repercussions on the system as a whole, and treating them in isolation will not reap success. The systems approach, which is somewhat akin to holism, is the obvious choice, as it is the most flexible: it offers no general guiding rules, as such rules do not help; it offers only partial rules, including those based on some biological or systemic or cybernetic theories. These may at times partly guide choice--in theoretical considerations as well as in practical affairs--but the system should remain open-ended all the way.

The limitation of a fragmented approach to medicine and the vexation it engenders are rather obvious: side-effects of treatment in one special area can occur in another and may thereby be neglected or be tended to by other specialists whose work naturally complements the work of the initial specialist, thus forming a whole system of treatment. Advanced specialization renders the system of medical practice quite complex and thus not easily comprehensible and not very stable. A new integrative attitude is therefore urgently needed.

Thus far, only the need to have an integrative picture has been noticed. This is too little for comfort. Specializations develop even when integration is effected. Systems theorists hope to obtain better integrations by constant attention to interaction while neglecting details when possible. The neglect may turn out to be objectionable; and then detailed, highly specialized studies lead to new integrative approaches and practices. For example, the discovery of the importance of hygiene in all branches of medicine, or the discovery of immune mechanisms, has challenged every branch of medicine. This is how different branches and sub-specialties of the medical profession often interact. The interaction is strongest and least avoidable when it bears on treatment. Complex interactions within the large-scale medical systems of our day require some forethought and planning and call for a commitment to the clinical perspective. This book offers a way of using both techniques, integration and specialization, either individually, when feasible, or together, as a new and highly specific high-tech tool which may serve as a means to forge a new integrative specialty. Utilizing the specific to enhance the integrative is a feasible and often practiced technique. Hygiene and immunology are not the only examples of this; some tools may serve just as well. A tool serving today an integrative function is the computer; the specialty is the study of integrative computer-assisted diagnosis; and the means is the proposed comprehensive computer-assisted medical diagnostic service.

The computer has diverse uses in medical practice. It is used as an aid in medical administration, in medical industries, and in both pure and applied research. It is used to broaden diagnosticians' field of vision and is currently applied to enhance the process of reasoning and decision making concerning matters diagnostic. Only a few decades ago, this was met with great hostility; now it is generally accepted, in principle, even though its utilization is generally limited to emergencies and to serious cases.

The special attention which the computer receives here stems from three sources: moral, methodological, and practical. The first, moral concern, is with those regrettably common diagnostic practices which restrict the freedom of patients as they are intended to rob patients of their responsibilities and transfer them to experts. A reform movement equipped with computer power should help remedy this. The second, methodological concern, is with precision. The introduction of computer-assisted mode of thinking will help raise the standard of the whole discipline tremendously. The third, practical concern, is that computer service in diagnosis is regularly introduced *ad hoc*, with no guidelines, quite contrary to tradition. Guidelines must safeguard the public against the irresponsible use of new tools and encourage their proper use. Without such guidelines the computer may inscribe irresponsible practices into hardware (since the difference between hardware and software is a matter of degree and not of any principle; designers feed into the harder parts of the system's program the aspects of a system that look to them more durable). A computer revolution sensitive to the medical, epistemological and ethical context will, hopefully, prevent this.

In addition to the theoretical and methodological aspects of medical diagnosis, this book is concerned largely with medical ethics, especially public ethical problems raised by modern science and technology. Moral problems cannot be studied in the abstract: they need a framework to serve as their general context. To that end this book offers a philosophical framework as well as general theories from different fields. This philosophical framework constitutes a new version of classical individualist liberal social and political philosophy and a new liberal philosophy of science; the new liberalism attempts to replace the outdated and impractical demand for proof with the modern demands for responsible conduct.

We address this book to medical practitioners, computer experts, researchers, educators and administrators, as well as to students of medical ethics--but mainly to medical students. This book is intended to serve not only the highly specialized professional reader. It is presented in a common language with little presumption of the reader's prior knowledge of the field, in the hope to assist the critically-minded readers who may wish to criticize us; we invite them to try to supersede us. The sooner this book will be superseded, the better off diagnostics will be.

New Haven, September 1987, Jewish New Year.

Chapter 1

INTRODUCTION

1. NEW TRENDS IN DIAGNOSIS

This *Introduction* includes four sections. In the *first* we present a general theory of diagnosis, exploring its limitations and its potential application to a comprehensive computer-assisted medical diagnostic service. In the *second* section we pose the question: Given the two conflicting ways of approaching problems, the integrative and the isolationist, How is a balance between them best reached? Otherwise asked, How can contextualization and idealization best be balanced? In the *third* section we discuss the guidelines for the use of computers in medicine and for social control over its use. We also reveal as futile the tendency to use computers as storage for all available data (in medicine as elsewhere). In the *fourth* and final section we pose an overview of the comprehensive computer-assisted medical diagnostic service of the kind here envisaged.

In each section, two views of science are contrasted, the traditional outdated view according to which science is inductively based on vast collections of facts; and the view which is nowadays gaining popularity, according to which science is hypothetico-deductive: science is primarily a series of theories, or of sets of hypotheses; it is from such a set of hypotheses that both known (repeatable) factual information and new (repeatable) information is deducible; and some of the new information is empirically testable. The *Introduction* ends with a summary of this volume.

- 1 -

2. AN INTEGRATED OUTLOOK ON DIAGNOSIS

In this section, we explore the need for a general theory of diagnosis and for one which includes an explication of its own limits. We shall call this "diagnostics". Let us first contrast the traditional approach to diagnostics with the latest proposal which may, with proper planning and reasonable effort, be implemented in the near future within an integrative and comprehensive computer-assisted medical diagnostic service.

Since modern medical diagnosis is couched within science, views on scientific method have naturally influenced the approach to diagnostics. To begin with views on science and its methods, the most important disagreement in methodology concerns the status of theories when they first appear. The leading traditional view was inductivist: new theories emerge from factual information, and when this information is broad and rich, then the theory based on it is scientific, proven, true and unquestionable. The more modern view is hypothetico-deductivist: new theories are sheer guesses; when they explain known factual information they merit tests by appeal to new factual information. The holders of the modern view consider the more traditional view a myth; the holders of the more traditional view consider the more modern view dangerous since people can easily find confirmations to their hypotheses, to their own subjective opinions, by stressing information that agrees with them and by overlooking or slightly distorting information which does not. Since the temptation to do so is too great, it should be checked, as its will make its holders dogmatists.

Despite the popularity of the view that the habit of trying to make hypotheses leads to dogmatism, scientists and philosophers of science repeatedly recommend this habit. They gloss over this inconsistency by wavering between different versions of the traditional theory. They shift back and forth between the view that hypotheses do cause dogmatism and the view that hypotheses may cause dogmatism. The first is trivially false: many thinkers repeatedly attempted to expound hypotheses and were not dogmatic. The second is trivially true: many dogmatists are captives of their hypotheses. The holders of the more traditional views also shift back and forth between the view that hypotheses are always forbidden and the view that at times the wealth of empirical information permits the making of a humble hypothesis. An explicit presentation of these views can rarely be found in the philosophical professional literature which airs them; rather, these views are adumbrated by the preoccupation with the assessment of the risk of offering unfounded hypotheses, as if it were irresponsible to do so.

The inner tension between forbidding and welcoming new hypotheses has risen considerably in the twentieth century. Due to the impact of Albert Einstein, nowadays philosophers of science almost universally endorse the theory of the empirical nature of empirical science as tests of explanatory hypotheses. The more traditional view that theory emerges from facts, nevertheless, still prevails. For example, it is taken for granted by many medical curriculum organizers. In their view, medicine begins with symptoms, from which it proceeds to etiology and then to treatment proposals. This traditional view of medicine encourages another traditional (and, we shall argue, unacceptable) view, still supported by the diagnostic literature, namely that diagnosis is limited to the initial clinical diagnostic encounter and is performed in utter disregard of any non-medical considerations, even the obviously relevant ones, such as the scarcity of available resources. The culmination of the traditional view is the demand of diagnosticians to be as empirically thorough as possible and to eschew more abstract and complex considerations. This model of conduct is regrettably sanctified by most of the standard diagnostic literature.

This traditional view presents science as a hierarchy, with factual basis and levels of theoretical superstructure; the base of medical science is then anatomy and the superstructure is first physiology and then nosology (the theory of disease); the base medical practice is symptoms, and conclusions about treatment should follow from their description in the light of medical science. The main prerequisite from science, according to traditional inductivism, is that the theoretical superstructure should emerge in steps out of its factual base and that prescription should be based on well-founded theory. This idea will be shown very influential, false and harmful; it requires that the diagnostic process should continue well into the stage of research and away from the means available to patients in practical situations.

The traditional view of the natural hierarchy of science is too neat. It is the ideal of science, not its reality. For real situations systems theory is more practicable; an outline of it will be presented in this book. The main corollary of it for the present discussion, however, is simple. In one situation **a** is a sub-system of **b** and in another situation this relation is reversed: there is no "natural" hierarchy. In this book we will illustrate this repeatedly and explain the need for the freedom to alter hierarchic relations as required, in order to construct the comprehensive computer-assisted medical diagnostic service which is envisaged here, as one which is the best implementation of an integrative approach to medicine. By "integrative", we mean "systemic", "systems-analytic", "comprehensive" and "synoptic". We discuss comprehensiveness or integration as exhibited in proper diagnostic use, well within the diagnostic process. The integration is of diverse factors, from therapeutics to economics; it shows in the diagnostician's freedom to take one set of considerations once as more basic to another, and once in the reverse order.

Much of the process of medical scientific and technological innovation is interlocked with the process of specialization, which has fragmentation as an undesirable side-effect. Medicine did not escape the threat of this side-effect. To counter this threat much integrative work is regularly required. One example of specialization is the recent introduction and development of computer expert-systems software, which will be discussed later on in some detail. Its ability to assist in diagnosis comprises a significant step forward on the frontiers of science and of technology, yet it also threatens medicine with further fragmentation. Consider the expert system that enables cardiac patients to use a telephone to plug their data straight into a remote computer so as to receive immediately instructions on their video screens. This system seems to be closed and interactive: the patient and the computer seemingly interact to the exclusion of practically the whole world. But the opposite is the case: computers are excellent means for communication, education and the broadening of horizons. The expert system meant to assist cardiac patients is not meant to increase their isolation, but to facilitate their self-diagnosis (especially in emergency) and their subsequent interaction with their physicians. Nevertheless, isolation is obviously a possible outcome of this innovation and so steps should be taken to prevent it. The present book depicts computer-assisted diagnosis as a highly specialized technology which can serve an integrative function and prevent the fragmentation of any medical service; we propose that this is best done in the rational design of the diagnostic process. Diagnostics itself, in its turn, may be enriched by this new and sophisticated technology.

Medical diagnosis is the concern not only of medical researchers and practitioners but also of patients, both actual and potential. Early detection of disease in seemingly healthy people is a powerful means of preventive medicine. Early correct or incorrect diagnosis is often what makes the crucial difference between success and failure. What often stands between a patient and the best available treatment is just the difficulties encountered before correct diagnosis is made, and the loss of irretrievable time, a loss that irreversibly prevents early and effective treatment.

Diagnosis is the weakest link in the therapeutic chain joining patients and health. The most efficient improvement of standards of medical practice is the improvement of medical diagnosis. Many believe that the weakness of current diagnostic practice could be both diagnosed and treated by the computer; this book recommends instead live diagnosticians who can effectively be assisted by many instruments, including computers and medical diagnostic expert systems.

There are a number of fields, from medical diagnosis to town planning, in which public opinion--expert as well as inexpert--holds fast to the conviction that knowledge explosion makes the use of the computer increasingly unavoidable. This is a claim which offers a diagnosis and a treatment in an easy package deal: the cause of our troubles is declared to be knowledge explosion, and the best treatment is declared to be computerization. This may be true, yet all easy package deals are suspect and require understanding before implementation. The very availability of a treatment, and the very success of that treatment in some area or another, give the impression that we are exempt from examining carefully the diagnosis both of the problem and of the claim that the readily available solution for it is also the best. This impression is very seductive as it quickly alleviates the strain of an urgent practical problem and as it promotes a very comforting diagnosis of the trouble as a mere by-product of a positive process: the growth of knowledge. This growth, considered essentially good, also causes some difficulties in its mastery, but with the use of computers it will be mastered soon and so there is no cause for worry. All this invites careful examination. What exactly is the problem of knowledge explosion? How will the computer help overcome it?

Computers have led to some invaluable successes in medicine as diagnostic tools: for example, computer tomography and medical expert-systems software programs. Medical diagnostic practice has been recently augmented by computer technology to the extent that it becomes impossible to discuss the one without the other. More important for our purposes, to enlist the computer's services in a proper manner one needs an explicit expression of theory, rules and principles of diagnostic practice. This explicit expression is our goal, as we see in it the means for the solution of problems which the introduction of the computer into medical diagnosis has made fairly common knowledge. As increasing numbers of competent teams of computer technicians and diagnosticians are now busy developing programs for computerized diagnostic services and for computer-assisted diagnostic services, in this book we focus on problems which this development brings about, and we seek solutions in the explicit statement of the most general aspects of the diagnostic process.

Consider the fully computerized diagnostic system, in one huge program which would include a complete list of known defects and complaints and deficiencies, a complete list of known diseases, and all the known correlations between groups of items from these two lists. This computer program can be obtained in a process similar to that of transferring a whole huge dictionary to a computer program in order to facilitate its use. Such a program would involve the transfer of the most comprehensive medical textbook to the computer system in order to facilitate its use. In principle there is no difference between computer programs comprising a dictionary, a workshop inventory or a medical textbook: they are all means for facilitating and regulating routine procedures. Yet in practice this does not hold equally well in all cases: in some case, such as that of diagnostics, the program may easily

overwhelm its programmer and its user by the enormous wealth
of possibilities it is meant to handle. In the case of chess, in
principle the procedure for creating such a program is the most
straight forward and unproblematic, yet in practice the mere
wealth of material, the imposingly large number of possibilities,
makes the program utterly impossible. How is the obstacle of
excess information overcome?

Consider the example of an inventory. In large
workshops workers follow certain routines and the plan to
introduce a computer to assist them in some of their work
follows that routine. A minimal amount of initial information
about some leading routine is used to generate a list of possible
hypotheses as to the items which may be needed while
following it. The hypotheses which are both most frequently
true in the routines in question, and the easiest to test, are
chosen to be put in the inventory as the cheapest ones to test
first: the product of the frequency of the truth of a hypothesis
and the benefit of the choice of that hypothesis when true is
its utility function; the choice which the computer program is
supposed to recommend is of the hypothesis whose utility value
is maximal and whose cost of testing is minimal. How cost
and benefit are balanced is a difficult question, handled by
cost-benefit considerations.

This is a general principle, and in diagnostics it creates
special and difficult problems. It is practically impossible to
create a computer system as good as can be done in principle.
And so systems are often built so that all the defects from
which they suffer and which can be eliminated in principle but
not in practice are easy to detect and easy to rectify. Fuses
are meant to function in this way: often replacing a burnt
fuse is all the repair a damaged system requires. The fuse box
is constructed to be the locus of trouble so as to be the first
place to examine: this way one exhibits one's first choice
among the available hypotheses to be that the fault is a burnt
fuse, though this hypothesis is not always true. Repair
technicians, electricians, mechanics and medical practitioners do
have some hypotheses like this one, even though it is not easy
to construct fuses for the human machine (except as
instructions: learn the signs of cancer and upon finding one go
for a check-up).

Take the two extreme cases of diagnostic problems (medical or not), the most easily soluble and the insoluble. In the first case support of the most obvious hypothesis is found with relative ease. In the other case practitioners have refuted all the hypotheses they could find, say, about the source of a fever; it is known as an F.U.O., a fever of unknown origin. Too often one hears a promise: in principle all problems are soluble; in principle a full list of possible hypotheses is available, so that one of them is bound to be true; in practice the list is admittedly too long to handle, but this will soon be taken care of. This promise should be criticized, because it is based on a dangerously false principle, one known as the principle of eliminative induction. The promise of eliminative induction is not only that certain formidable obstacles will be removed, but also that afterwards complete success is assured. Complete success, however, is never assured--even given enough time and resources, which, of course, is never the case in actual medical diagnostic practice. Nevertheless, it is of extreme practical importance to observe that practitioners, whether mechanics or physicians, do use a method akin to eliminative induction, even though never fully assured of success. Cost-effectiveness considerations recommend the use of limited lists, which are handier than the fullest available ones. There are successively richer lists, but even the best is not complete: the fullest possiblelist of known hypotheses may be exhausted and a diagnosis may still lack a successful completion.

The principle of eliminative induction is objectionable not because of its false assurance of success but because it distracts one from the task at hand: the principle is meant to constitute an incentive for research in the direction of seeking an ever increasing list of possible causes of diseases, whereas research has to be concentrated on the worthiest search, whether for a new diagnosis of a hitherto undiagnosed symptom or disease or for the way to systematize the well-known need to create efficiently series of successively detailed partial lists of possible diagnoses.

These successively detailed lists can easily be put into the computer memory. The idea of a series of lists engulfing each other like onion peels has been used in compiling dictionaries and inventories; they were taken for granted in computer technology since its very inception. Yet medical diagnostic information still has to be processed in order to be listed in computer programs in this, fairly familiar fashion. The rule for the processing or generating the lists is logically prior to them, yet it has been widely ignored even by the very people who attempt to generate them, not to mention the practitioners who apply lists of this sort which they keep in their expertise fairly intuitively. This makes the keeping of the lists rather uncritical and with too little updating. This is one of the obvious and serious gaps in medical diagnostics which, we hope, this book will help to correct.

Our final technical point about the successive sets of competing hypotheses is that they mark levels of expertise. For the lowest rung, the first set will do. If it is exhausted without the diagnosis being successfully concluded, the next level of expertise is applied; and so on. How are these sets arranged, where is each possible hypothesis located, and what is the general situation which emerges when such a system is implemented? In each field these questions invite a study of the situation in general. In the field of medical diagnosis the situation has hardly been recognized and so there is hardly any study of these questions. Why?

The traditional neglect of diagnostics lies in the structure of the traditional diagnostic literature. The most advanced medical textbook neglects entirely the field of diagnostics. The impact of such neglect is enormous: most physicians are blind to the basic principles that guide their own daily clinical activities and their own most basic science. They are, therefore, devoid of high level control over the treatment they offer. Consequently, they cannot inform patients rationally about some of the most basic aspects of medical practice. Consequently the recent growth of consensus on both the moral and the practical value of patient informed consent for medical treatment has not yet been accompanied with any growth of the practice of exercising it properly. This is a blockage often due to physicians' own limited knowledge of diagnostics. This limitation invites diagnosis.

The constraints on the options open to patients, and their diverse possible attitudes, constitute familiar significant factors. Even moderately competent and moderately experienced diagnosticians involve their patients in the diagnostic process by discussing these factors with them during the diagnostic encounter; they take as self understood the relevance of their patients' responses to constraints outside the field of medicine, such as constraints on their budgets, time budgets, freedom of movement and travel, and commitment to family and work. Experience acquired in such encounters is not pooled: conclusions are thus often reached less efficiently than they would be were options carefully calculated and compared. To this end, however, the myth that the diagnostic encounter is confined to one preliminary session should be destroyed, so that both patient and diagnostician may feel free to admit ignorance and agree to meet again after a brief while. At times this is done--when patients wish to consult relatives and when physicians find it difficult to break bad news suddenly or when they wish to call a conference. These simple solutions are prevalent and yet they have not destroyed the popularity of the myth that diagnosis is confined to one preliminary clinical session.

Medical practitioners are required to account rationally for their diagnostic activities and yet, wanting perspective on them, they offer increasingly unsatisfactory accounts and get lost in technical details that fail to amount to any overall picture. Consequently, physicians and patients alike are threatened with total loss of control. This book is aimed at explicating the problem within its most general context--i.e., the theory of medical diagnosis. We aim to offer a tool for the acquisition of overall pictures which will help fill a theoretical lacuna and facilitate the overall improvement of medical services, for individual patients, for their physicians, and for society at large.

The trouble with current practices of medical diagnosis is the following. In general terms, the difficulty lies with some of the present methods of diagnosis. For example, those diagnoses which do not fit the traditional myth of treatment as emerging from diagnosis (traditional inductivism) are seldom taken to be a refutation of that myth: rather, they are excluded from the field of diagnostics. Monitoring, for example, is rarely taken as diagnostic, though it clearly is; its being intertwined with treatment refutes the traditional myth unless it is not considered diagnostic proper. The disregard for monitoring as diagnostic, however, encourages approaching those diagnostic matters which interlard with treatment too subjectively, while encumbering considerations pertaining to them with masses of irrelevant details. Consequently, results can scarcely be tested by either physician or patient. They can be checked by public authorities or by researchers, but this is done too seldom, though some recent improvements have come about in this manner.

The neglect of the general theory of medical diagnosis is reflected in the fact that most of the diagnostic literature is specific, and is usually applied with little regard for diagnostics or for any other principles of applied medical science. Combining the fruits of modern research and of computer technology should be involved in the process of reform. Medical research often offers discoveries that improve both diagnostic techniques and diagnosis proper, i.e., the ability for and the knowledge of differentiating diseases and of pinning one of them down. Specific programs for computer-assisted diagnoses effect similar improvements, especially where the computer may be deemed as a part of a sophisticated diagnostic tool, since alongside the stethoscope, for instance, diagnosticians employ computerized axial tomography (CAT) scanners. Also, computer programs may aid the thinking process of differential diagnosis, particularly diagnostic expert-systems software, as in the programs that aid diagnosing infectious diseases. Yet, the evolution of computer-assisted diagnosis is unplanned, and not always directed at solving the general problems or the most urgent problems of current medical practice. Whatever exactly is the current state of the art, we are not familiar with any general discussion of the scope and limits of advisable computer service in medical diagnosis that takes the general theory of diagnosis as its context. This is what we venture to explore here.

The relevance of computers to all technology goes deeper than the mere usefulness of the services of computer technology. To see this we may consider, first, the application of that service to the limit of the potential of diagnostic technology. Consider any part of diagnostics as possibly open to improvement through the use of computers. Assume that at a given stage of diagnostic knowledge all that it includes that can be relegated to the computer has been computerized, so that the rest invites a human, non-computerized complement. In this situation the ambitious outline of a comprehensive computer-assisted medical diagnostic service is recommended--to be used until further knowledge is acquired. Can this situation be reached? If so, should it be reached? Today this can be done only with the support of governments. Yet doing nothing until governments endorse it may be very harmful. Instead, the general idea of a comprehensive computer-assisted medical diagnostic service may be used at once as the intellectual framework, as a regulative idea; the regulative end-point is a computer network linked to a central service subject to public control and regulation.

Proceeding this way may yield techniques for discovering the most efficient use of available computer power. Proceeding this way also amounts to a presentation of the general diagnostic situation as formally and explicitly as possible, thus offering current ideas in a manner most open to criticism and to subsequent improvement. This is why the system of medical diagnosis is better examined within the context of the computer revolution and the latest computer revolution within the most general theory of medical diagnosis. Though the comprehensive computer-assisted medical diagnostic service, when properly implemented, will revolutionize medical practice, the immediate recommendation should invite partial implementations of the reform here envisaged, and only in an experimental way. The reform in the public awareness to the problem has to be thorough: diagnostics ought to be put on the agenda of any responsible medical diagnostician.

3. THE DIAGNOSTIC PROCESS IN ISOLATION

Diagnosis concerns the knowledge and assessment of the patient's complaint. It differs from other parts of medicine, such as etiology, which concerns the cause of the complaint; and from prognosis, which offers conjectures about the likely course of the disease; and from treatment. Admittedly, it is intimately connected with these parts of medicine. Thus, physicians rarely diagnose beyond the limits of treatment; they use treatment as a test for the diagnosis and, at times, as a diagnostic tool. They test the diagnosis by prognosis, and they test diagnosis by etiological (e.g., pathological) examinations.

Though the use of treatment and prognosis as diagnostic tools is well known, even to the extent that there are different ways of using them, the lack of explicit statement of the matter prevents progress. There are cases of choice between a specific treatment that is a good diagnostic tool and a broad-spectrum treatment that is often safer but always less informative. (The broad-spectrum alternative is dangerous, e.g., when it may worsen some possible infection). In medical schools the frequency of the specific treatment is expected to be above average; the broad-spectrum treatment is more frequent in field conditions where the chance of a patient returning to the clinic are small. The fact that there is no articulation of all this in the literature precludes carefully detailed statistical hypotheses and their testing and implementation: the field still is a *terra incognita*.

Despite the evidence that diagnosis and treatment are interlarded, diagnosis is a special and separate part of medicine. Without it, empirically-based medicine and empirically founded specific treatment would be inconceivable. In particular, patients may want a break between diagnosis and treatment so as to come to some decision, and physicians then often concur. This interruption comes to secure the patient's right to decide on which the demand is based that treatment be conducted with informed consent.

Conflicting tendencies come into play here--one integrative, the other isolationist. Some guidelines are required, preferably derived from common sense, to demarcate and help balance the conflicting tendencies. There are some obvious integrative questions that are relevant to diagnostic practice, which were considered external to it until recently, such as the question, What proportion of the population should join which profession or specialty? These questions were customarily left for the internal regulation of the various professional associations, since they did not constitute a pressing concern before the advent of the population explosion after World War II and the new global mobility that began roughly at the same time. Today the need for guidelines is generally recognized: How much should the diagnostic process continue in isolation, based upon internal criteria, and how much should it be directed by external considerations?

The simplest proposal is as follows. Consider diagnosis in isolation but explore each aspect in a tentative, open-ended manner, allowing the introduction of a more integrative view when appropriate.   Thus, at any given point, external factors, such as the cost of treatment, and even the cost of further diagnosis, may enter deliberation.   When the patient is insured, for example, this deliberation may be postponed.   But it may surface later, when insurance runs out.   In other cases, the deliberation on insurance will intervene earlier, on the personal level, as when a person deliberates the choice between various medical insurance plans available on the market, or on the national level, when the government considers its budget.

This proposal seems commonsense;   there is no need to argue that it should be accepted, since it is not contested. Deviations from it may thus be considered plain errors.   Yet this common practice should be checked and controlled to safeguard the public interest.   How can this be accomplished?

First, one may notice simple and clear examples of oversight, such as those which occur every time a strong interaction between diagnostics and other aspects of medicine forcefully come to fore, the delay in noticing it has gone too far.   It is better to institute safeguards in advance so that the complaint will be voiced before the penalties become prohibitive.   If diagnosis is insistently taken in isolation, then eventually an impasse will be reached, and it will forcefully impose extra-diagnostic factors in an unplanned *ad hoc* manner. Rational planning would have been preferable.

In principle, physicians are ready to perform diagnostic services for any conceivable length of time.   But they often find themselves limited by practical considerations such as the patient's or insurer's ability to cover costs of initial diagnosis, especially when expensive procedures or lengthy hospitalization are involved or when there is a shortage of instruments or beds.   The fact that diagnosis can be dependent on these issues invites the establishment of policies regarding them.   Moreover, while increased affluence can help render a procedure affordable, it can also paradoxically give rise to newer and yet more expensive diagnostic and therapeutic techniques.   This situation can both make more sophisticated procedures available to those who can afford them while increasing the number of cases for which other issues will significantly influence a given person's likelihood to receive the best diagnostic service available.

The current situation is immeasurably improved in the affluent parts of the world.   Yet medical services there are now suffering from a severe depletion of financial resources which is partially caused by the absence of rational budgeting and computations of cost effectiveness on a national scale. Planners of national medical care and of other budgets for medical care look for guidance from diagnostics and epidemiology. The paradox of higher medical expenses due to increased affluence also expresses itself in the unplanned increase in the supply of the highly specialized services in response to a demand that has not been studied as yet. Patients entering highly specialized clinics, for example, may receive inferior diagnostic service or treatment, unless the service were enlisted of the right specialist, the one who happens to be best qualified to diagnose or to treat them. This well-known fact offers the rationale behind some of the best medical education services, where practitioners are trained in the art of choosing specialists for both diagnosis and treatment (since such decisions require preliminary diagnoses, of course). But even these programs with integrative aims, need explicit, well-tested guidelines to help determine, at the very least, the proper balance between the isolated and the integrative approaches.

This proposal, to develop guidelines for the proper balance between the isolated and the integrative diagnostic approach, clashes with the traditional and still popular idea of all diagnosis as isolationist. Yet all practitioners know that perfect isolation is impossible and it would seem they would therefore welcome such guidelines. Yet these guidelines are extremely hard to generate: the integrative aspect of diagnosis is often very far-reaching and all-embracing: often the deviation from isolationism brings in a whole complex system of broad considerations which depend on whole lifestyles, especially where chronic illness and convalescence are concerned. Detailed systems analysis may render these considerations manageable by breaking the system involved into sub-systems. This matter will be treated after discussing some system-theory considerations, especially cost-effectiveness considerations.

The criterion for drawing the line between the isolationist and the integrative approaches is, very roughly, this: integrative considerations should be omitted when they become either too problematic or too marginal. This rough criterion is operable when small-scale cost-effectiveness considerations are worked out and computed in sufficient detail and accuracy.

Yet without the aid of small-scale diagnostic cost-effectiveness computations, the outcome of such an exercise may be very misleading. Even when cost-effectiveness considerations are applied to matters of the national budget, where it is possible to base them on given epidemiological information within budgetary constraints, this is not what is done: matters of lifestyle enter this case too. Financial considerations applied to national statistics may suffice for the computation of national budgets, yet added factors normally enter as some weight functions, which are means of expressing numerically certain preferences. Certain services may receive preferential treatment in order to insure that they will not disappear completely: when the budget for a given service (or a given style of practice of a particular service) becomes so small that its existence is threatened, it will be ascribed an increasingly high weight function to insure its survival. Similarly, weight-functions may insure a minimally reasonable distribution of a service in different areas, different social classes, etc. Whether or not the weight functions are usefully applied and correctly estimated, it is simply impossible to convert small budgets to national budgets without putting them to some use: they constitute the bridge needed to convert small budgets and population statistics to national budget statistics. They are indispensable even in large medical service complexes, such as large university hospitals.

It should be noted, perhaps, that the choice of weight functions is always problematic. Some of the problems involved may be better tackled when certain factors are explicitly recognized: the methods of medical diagnosis, the concerns over practice variation, the limits of total medical expenditures, moral considerations as to informed consent and confidentiality, and so on. The current situation on these matters is still largely unstudied; yet it is legitimized and perpetuated through practices that are engendered *ad hoc*. The study of the economics of public health is in danger of becoming extremely biased by the high incentive that supports and legitimizes current practices merely because they are current, and is thus a disincentive for any progressive change.

Still, recent public health changes in the United States, driven by economic thinking, did encourage some radical modification in the style of medical practice. Their aim is to discourage expensive medical procedures and encourage evaluation and management; to discourage sub-specialization; to discourage hospital-based practice; and to encourage rural practice. Immense variations in practice and high medical costs invite the suspicion that there is excess utilization of medical practice, namely, over-diagnosis and over-treatment. The suspicion was backed by economic considerations, aided by statistical computations, without proper consideration of weight-functions, often of financial cost with no attention to medical benefits. Though economic considerations are highly called for at this juncture, such practices are irresponsible for their disregard for many factors (including quality of care) that constitute medical diagnosis and treatment. These practices can be improved by better use of diagnostics as well as of control, and both are amenable to the service of the computer.

The economic aspect of the matter should be simple, as the incentives it relates to are financial. Yet other incentives exist, and current practices always invite attempts to legitimate them, so that linking legitimation and incentives always tends to impede progress. Moreover, legitimation and incentives are always linked unless there are safeguards to prevent this linkage. Even so, the situation is still scarcely studied. The likelihood of progressive change is unknown. Nor can one study its probable causes. Yet, if a change is called for, and if it is desired that change on the national level be efficiently implemented, then it is not difficult to see what might be the best starting point. The division of populations to diverse diagnostic sub-groups is crucial for the allocation of the budgeting funds. This division is based on some accepted diagnostic and therapeutic classification, which may be subject to study and revision. The study would depend on the endorsement of certain values and the development of some general theory of diagnosis. Thus, the starting point is diagnostics. In the absence of a general diagnostic theory some intuitive diagnostic classification is used. Often the intuitive classification of diagnosis replicates the extant division of medicine--into specialties; it reflects the view of diagnosis as separate from treatment and entrenches the current situation.

The impediments to improvement are not due to any fault, certainly not any fault of medical practice, education, or research, but are the consequences of poor budgeting and poor planning--personal, communal and national. The poor planning on the national level has led affluent nations to restrain the financial allocation for both medical training and research. More rational planning might result from public discussions in the professional associations as well as in the national political institutions. Without democratic scrutiny and democratic control, planning is left to power brokers. We propose that the starting point for the public discussions concerning medical budgetary plans should be diagnostics. *Treatment without diagnosis is blind; diagnosis without treatment is lame; each of them provides the context for the other.* Hence, diagnosis is the chief tool for rational control over-treatment, and the budgeting allocation of medical funds is the responsible control over such medical activities.

The intended beneficiaries, we propose, are the patients; yet the immediate beneficiaries might very well be medical planners and administrators. This profession and the problem just described are new; considering the proposals here voiced favorably or not, as ultimately judged objectionable or acceptable, may clear the ground for a more comprehensive policy.

This, then, is the limitation on the proposal made here: **to keep the diagnostic encounter isolated when this can be done comfortably and responsibly, but to keep it open-ended in any case.** The proposal requires global considerations to render it possible; guidelines for when the diagnostic encounter cannot be taken in isolation have to be agreed upon. At times this is extremely hard. An extreme example is political torture. The traditional guideline for physicians was to ignore all politics. Today this guideline is questioned in case of political torture, since it leads to the view of the recalcitrant physician as an accessory to a crime against humanity. Whatever the guideline should be, it now has to be clearly stated and aired in public discussion. This is not the only political situation impinging on medicine, though regrettably it is one of the most topical.

This should suffice as the preliminary discussion concerning the two conflicting tendencies, in discussions of medicine: the tendency towards isolating the problem at hand and the tendency towards developing an integrative view out of isolated pictures. Both tendencies are of extreme value individually, but using them together requires care.

The integrative tendency is to discuss matters in their broadest context and to make them look as realistic as possible; this tendency must be checked by its opposite, the tendency is to isolate, to abstract cases from their context artificially in attempts to generalize them and reach interesting insights; but this tendency often leads physicians to the notorious view of patients as mere cases and must thus be humanized. These two tendencies are also known as contextualization and idealization; each of them has led to enormous achievements though neither can have any guarantee of success. Moreover, when the two conflict rather than complement each other, there is even a guarantee for failure. It is also advisable to remember that both contextualization and idealization may complement each other in the service of both good and bad causes, such as the evasion of criticism and thus the prevention of innovation. Evading criticism is defensive and counter-productive, as when an idealization or a contextualization of a particular diagnosis is proposed or rejected.

We propose that contextualization and idealization should be strictly tested both for their usefulness and the correctness of outcomes. The outcome expected in this study is an integrative though open-ended framework for diagnostics--to be embedded later on within the general context of the more general theory of medical practice.

4. SUMMARY OF THE PRESENT BOOK

Chapter Two.

The canonic diagnostic literature includes the set of rules that diagnostic deliberations have to follow, and the diagnostic context is medical science and the relevant social customs and mores. This context is now enriched by the entrance of the computer into medical service. The diagnostic text, therefore, should be enriched by additional rules concerning three kinds of deliberations, first each separately and then all three in conjunction. The guiding question concerning computers in diagnosis is, How far is the formalization or the computerization of the diagnostic process (a) permissible, (b) feasible, and (c) beneficial with no discordant ill side-effects?

Chapter Three.

The computer revolution in medicine has already quietly occurred. It has been introduced with no planning and on the basis of the erroneous standard views of both diagnosis and methodology: the standard view of diagnosis as limited to initial encounters, which has prevented critical discussion of the broader context of diagnosis, and the standard view or rather myth of science as based on a lot of unordered data, which has blocked critical discussion as unnecessary. Yet the advantages are tremendous and new and exciting and involved with fearful risks. The great overall advantage of the computer is its ability to increase precision and reach utter formality. The limits of its usefulness lie in the limits of these advantages.

Chapter Four.

Individualist ethics is taken for granted here and viewed as embedded in the social system. It requires informed consent not only in treatment but also in diagnosis. Introducing informed consent into diagnosis will convert it from a litigious principle to a prescription for cooperation. Informed consent raises problems, both moral and practical. The implementation of a moral principle requires awareness of practical constraints: the actual conditions and the revolutions in medical practice. Moreover, it requires that scientific research be viewed as a private activity and its products as a public good, especially when technologically significant. Products of technological research are viewed differently--by law and custom. Patients are both consumers and guinea-pigs of the diagnostic system; this may be maximized by a comprehensive computer-assisted medical diagnostic service.

Chapter Five

The classical mechanist approach now gives way to the systems approach. Its applications illustrate how deeply individual diagnostic sessions are context-dependent--with the whole system of national public health as their context, legal, social and economic, but also medical: the individual diagnostic sessions interact with national epidemiological statistics. The systems approach is problem-oriented and embraces a program which suggests methods of viewing specific items in different specific contexts. An item considered as text sine-context (i.e., without context, ideally or in the abstract) may be placed in different contexts, all of which belong to a para-text (i.e., a set of closely-related alternative contexts), and this ought to be done in accord to meta-textual deliberations. The meta-text should then prescribe different contexts for different problem-situations as the better approximate modes of given problems. A systems analysis of the systems approach itself reveals that economic and moral problems are always central to systems analysis, and that they are only soluble with regard to the goals of a given system's decision-making sub-system. Since the aim of diagnosis is cure, diagnostic and therapeutic considerations intertwine (and largely via cost-effectiveness computations). The aim of the medical diagnostic system is dual: clinical and research--both interacting strongly with public health. Hence the tension between the isolationist perspective as concrete, and the systemic perspective as abstract. The latter view allows formalization and standardization and is the basis for diagnostic expert-systems software services. Today these are naturally but not thoroughly systems-theoretically oriented. Their marketing should be socially regulated and their application to the concrete case needs to be refined by clinical considerations.

Chapter Six

Viewed as pure science, medicine has no patients and no known performance measure. It is therefore better viewed as a sub-system--a context--here, as the context of diagnostics. Diagnostic technology is more than applied science, as it includes economic, legal and moral factors. Computers may help make the considerations regarding these factors more explicit and more rational. Diagnostic methods are always questionable and open to improvement in order to make diagnostic technology more effective and more responsible.

Chapter Seven.

The traditional theory of medicine presents treatment as based on symptoms, syndromes, disease entities and etiologies--in that

order. This is a gross error, since diagnosis cannot proceed
without hypotheses which violate this order and which should
be tested. An example is the null hypothesis--namely, that a
given patient is well. Only the explicit introduction of
monitoring over the clinical diagnostic encounter can help
maintain its uniqueness as both subject to rules and open-
ended. The computer can be helpful in this regard, provided
its limit be recognized and complemented by responsible
humans.

Chapter Eight

Diagnosis as a process of pattern recognition overlaps with
formal information theory. What signals ought to be read as
carrying information? Can reading-errors be detected
responsibly? The diagnoses of single cases and of public health
serve as monitors for each other. Standards for responsible
error are socially regulated by licensing.

Chapter Nine.

Subjectivity can be formalized with ease but to no consequence.
The limitation inherent in any human system is that of
malingering: the use of the system against itself. This
limitation shows the impossibility of replacing all symptoms by
signs. This is particularly so in the case of functional disorders
which, in the present state of the art, limit the use of
computers in diagnosis. Reliability in science and in
technology, including diagnosis and computer technology, is
inherently linked to repeatability. Commonsense reliability may
be different but then it is beyond the scope of the present
volume.

Chapter Ten.

The expert-systems software programs for partial diagnosis
which are now on sale in the market are expandable and
capable of linkage with each other. Therefore they may be
considered the first stepping stone towards the comprehensive
computer-assisted medical diagnostic service repeatedly
advocated here. Though the idea of a comprehensive

computer-assisted diagnostic service may sound utopian, it can be implemented gradually while being very useful at once. Legislation, however, ought to regulate against the readiness to exceed the limits of automation. But even without the full implementation of a comprehensive service, the view of it as such can function as a regulative idea and should do so as soon as possible. This will prove useful as long as diagnostics at large remains within the established medical standards, while the standards themselves undergo monitoring by a public body with an eye on their improvement. To maximize utility while minimizing intervention, reform should center on improving current lifestyles, each in their own lights.

Chapter 2

THE PROBLEM SITUATION

AND ITS CONTEXT

1. THE PROBLEM

The problem to be discussed in this section is very straightforward; discussing it is not. When and to what extent are diagnosticians allowed to deviate from the rules of the diagnostic procedure? To examine this it is best to embed it within a context. This leads to another question, How is the context chosen to begin with? After answering this question we turn to determine the contexts to our initial problem, concerning the desirability, feasibility and permissibility of deviating from the established rules of procedure. We are then ready to move to the next section, on uncertainty, where we may begin to attempt to answer our initial problem.

The diagnostic process is governed by established rules, supposedly stated in the canonic diagnostic literature. The claim that a given population behaves according to clear, fixed formal rules is best tested by computer simulation. If the claim passes the test, then the computer adequately simulates humans and may be used to duplicate or replace them. Hence, if diagnosticians rigorously follow established rules, they could be replaced by computers. If not, they may be seen as taking liberties and experimenting with the lives of patients. Both these options seem inappropriate. Where is the error? It is this: by rigorously following the same established rules, experts may be prevented from doing worse than the average, but at the same time, they will be prevented from doing better than the average. Do they then deviate from themselves? How do diagnosticians perform either better or worse than the average? By deviating from the established norm, of course. What gives them license to deviate?

The seriousness of the problem may be better appreciated if the use of computers in diagnosis is considered a bit more critically. How can computer techniques be best applied to the service of medical diagnosis? In particular, What is the price required, and are prospective users able and willing to pay it? What modifications should the diagnostic practice undergo to accommodate the introduction of the computer? What are the moral constraints on this modification and this introduction and how should they be instituted and safeguarded?

The introduction of the computer in the service of diagnosis may be compared to its introduction in the office. Even opponents to computer diagnosis concede that the clinic receptionist's computer is quite unproblematic. Yet potential problems can be found in this context too: the computer may mishandle data; it may have adverse effects on the user's health; it may cause problems for public accountants unable to properly audit books stored in computer memory; and it has already raised problems of the protection of privacy, since computer-stored information can be stolen in many ways. All this need not be discussed here; these matters are explored in the literature. Yet in each field, **each new technology poses new solutions coupled with new problems**. The introduction of the computer in the (clinician's) office is far from being simple and raises a problem that has at least three facets: technical, economic and moral.

The context of the present discussion is the question, what the place do the established rules of diagnosis occupy in the diagnostic practice? The discussion shifts from these rules to the introduction of the computer into the diagnostic process, because the computer always follows strictly worded rules and cannot deviate from them. The computer may thus serve here either as a metaphor for rigor or as a real case-study of increased rigor. Yet, either way, one may be critical of the introduction of the computer into the discussion concerning the place of established rules in the diagnostic practice, since this way the description of the role of the computer in medical diagnosis tends to be somewhat idealized: technical, economic and ethical considerations are in; other considerations, for example, medical politics (of medical administrations and of medical associations), are out. There are general arguments in favor of the implementation of computer services in various scientific ventures for administrative, technological, or research purposes. Such arguments apply to the problem at hand, namely, to problems concerning the use of the computer in medical diagnosis. To repeat, this problem is concerned with the desirable scope of the technically available computer service, its economic feasibility, and its moral permissibility. These three aspects of the problem constitute the present problem-situation.

Why are other parts of the context ignored, such as medical politics? How can each of the three other components mentioned here be taken separately when they are so obviously interconnected? This leads to a brief discussion of contexts. This discussion can scarcely be avoided, since there is a need for a rule to balance the approach to any problem as idealized, as pure text, context-less, sine-context, in a manner opposed to the systems approach, to the approach to the same problem within a given context. We postpone this discussion for two paragraphs, one on terminology and one on the choice of contexts in general, namely on systems theory.

First terminology. One may refer to a problem or a statement in the abstract, as a pure text, as a text sine-context; one may refer to its problem-situation as its context. The para-text to a given text is then the context of the context, or the set of alternative contexts to the given one. Finally, there is the meta-text, the text that determines the boundaries and the nature of the context and of the para-text. It thereby also determines the problem or statement at hand. So much for terminology (*sine* is without; *con* is with or together; *para* is nearby; *meta* is beyond); its usefulness will soon be apparent.

Second, the choice of context. Meta-textual deliberation and decision is as tentative as any other. In one case of a given problem-situation it is better to consider economics in the context and politics in the para-text; in another case it is better to go the other way around. This is the crux of the matter: tentativeness. Before tentativeness was legitimized (by Albert Einstein and Karl Popper) the established choice was between utter idealization and its extreme opposite. Utter idealization is the view that the universe is built of absolutely separate bricks; this is the mechanist metaphysics. The opposite to idealization is the view that the universe is one strongly interconnected whole; this is holism, or holist metaphysics. It could scarcely be subject to critical examination and therefore was deemed mystic and obscurantist (and in an exaggerated move all mysticism was then declared obscurantist). Tentativeness opens the door to a middle way, to viewing some parts of the universe as more strongly interconnected than other parts. The view of scientific theories as approximations to the truth is essential here, since systems are obtained by overlooking some links and by permitting some idealizations which are distortions. At times the assumption of isolation is not an irredeemable distortion and can be taken as a good first approximation. In the second approximation one may add to the idealized case some item, by transferring it from the para-text to the context. One may try to add different items, alternatively or in succession, as the meta-text may tentatively prescribe. This brief presentation of systems theory may help elucidate the situation in diagnostics; the standard medical literature presents idealizations all the way, and without alerting readers to their pitfalls. In the present chapter three factors are recognized as the context of diagnosis--the technological, the economic and the moral--to the tentative exclusion of others, such as the political.

Here, then, is the answer to our own question. The exclusion of national politics and the politics within the medical profession, or rather leaving them in the para-text, is rooted in our view that they do not impinge on the problems which today beset the established practices of medical diagnosis as closely as the items we prefer to keep in the context of our discussion, namely the three factors, the technological, moral and economic. The political aspects we prefer to leave in the para-text--as matters to introduce only in expediency. We invite the reader to perform variations on our exercise. Similarly, bringing into context separately the three factors mentioned, the technological, moral, and economic, is superior to excluding them but at times it is inferior to considering their mutual influences--inferior, that is, in all cases where the interaction between them is significant. Thus, disregarding the moral factor amounts to saying that all action beneficial to patients' health are permissible. In a later chapter we discuss the moral constraints on medicine. Similarly, disregarding economic factors amounts to the suggestion that money is no object, that the patient is sufficiently wealthy. We discuss the economic constraints in a later chapter, on systems, since the economic deliberations concern cost effectiveness, which is best discussed in that context. There are, of course, situations where the economic and moral constraints impinge on each other. We shall have little occasion to discuss these and we leave these as exercises for the reader.

An example for the strong interaction of the technological and the economic considerations, however, is obvious enough to be stated immediately: economic considerations suggest that the service offering fully computerized diagnosis, though available in principle, would incur exorbitant cost. It is therefore preferable to confine our discussion to a service which offers much less, but in an economically feasible manner, one that offers the computer part of a computer-assisted medical diagnostic system. This might compensate for the decreasing supply of physicians and the increasing demand for them, thereby propagating a partial solution to the ever-pressing economic problem of providing adequate health care for the ever-increasing world population. The moral and the technological factors strongly interact in the very same manner, though perhaps in a less obvious manner, so that we shall have to discuss it later: fully computerized diagnosis, though available in principle, is quite immoral.

This most preliminary discussion of the economic and moral aspects of fully computerized diagnosis illustrates the fact that established rules are made in accord with decisions, and that these decisions are not very rigorous and not rigorously worded. Thus, one may observe that the presence of established rules in diagnostic practice is less problematic than it sounds, and that the freedom to deviate from them and the extent of permissible deviation belong to the system of established rules. Yet this observation raises questions such as, How are rules to be established? How are they agreed upon? To what degree of rigor should they bind? How much arbitrariness of deviation from the established rules is deemed permissible and why?

The economic considerations evidently enter the answers one may wish to give to these questions. But how? In matters of life and death, is it proper to limit diagnosis by financial considerations? It is evident that the problem of economic feasibility seems to be much less pressing than that of moral permission to use rigorous established rules--and thus the moral permission to use available computer technology. The moral problem is all the more pertinent, because new medical treatments raise new problems concerning matters controlling life and death. Also, since these new treatments are largely due to innovative mechanization and/or newly-acquired medical expertise, they raise the question, Where does responsibility for these decisions lie? Hence we should first discuss, What task are computers invited to perform in the domain of medical diagnosis? How well can they perform this task? And then, By what moral code and at what expense? We should thus discuss the technological, the economic and the moral aspect of medical innovations.

Limiting ourselves to an examination of the ethical considerations involved, we have said, amounts to ignoring the economic considerations and taking the technological ones as given, while idealizing (namely, without considering contexts). This leads to the following simple general consideration for the case of a hypothetical patient with an unlimited budget. How can this patient expect to get the best medical treatment available? There is an initial difficulty here, since even if our hypothetical patient can secure the services of the best specialists in the medical profession, the success of treatment is not guaranteed. Moreover, the service of the best experts will not necessarily be the best available, since even the greatest experts cannot be expected to be in perfect possession of all the up-to-date knowledge--not even in their own narrow field. Thus, of necessity, their diagnosis and treatment will be somewhat deficient relative to what is ideally the best available at the time. This is only one item which requires constant attention, and the situation can be polarized by the introduction of any computer service: the delay can be made unnecessarily long by delaying the updating of the information fed into the computer's data bank. In this respect the help which might be expected from computers in medicine is obvious, since computers are generally known to be an effective means for relaying new information fast, and since the general problem of updating information may be best handled by trained computer technologists. Still, every specialization, including medicine, presents particular problems, which require computer technologists in its service to possess some specialized knowledge. Thus considering the idealized we see where it breaks down and requires extension.

As computers can emulate any set of sufficiently rigorously worded rules, simplicity suggests the question concerning the idealized case involving the best experts serving the richest patients: How, if at all, will a computer help in that case? Looking at an idealized computerized medical care system with unlimited resources, where computerization enters as both a concrete case and a metaphor for rigor, we ask, What kind of use is there to the established rigorous rules of diagnosis and their rigorous application?

The most obvious service a computer may render is that of a handy catalogue or dictionary whose items are readily available by the push of a button. If medical diagnosis is viewed as a technical exercise, as a deductive process which could be computerized, then it matters little as far as the outcome of diagnosis is concerned whether computers are used in the process or not (except for the checking of errors, including faulty deductions). The question is then, Is the use of computers cost effective?

The problem becomes entirely different and more intriguing if the process of diagnosis is not purely deductive. Medical diagnosis is all too often claimed to involve some non-rational processes, termed "medical intuition". This brings back the initial question in a new wording: Should reliance on medical intuition be permitted or forbidden? If the diagnostic use of intuition is forbidden, then perhaps the way to computerize medical diagnosis should be found; if the use of intuition in diagnosis is permitted, then perhaps the way to control it should be found.

Medical intuition defies description but perhaps there are ways to describe some of its significant characteristics. If so, how might it be amenable to, or at least compatible with, the use of established rules and thus adaptable to computer use? We will explore these questions in later chapters. Here, we will explore the context for this discussion.

2. THE CONTEXT: UNCERTAINTY

Diagnosis is a process whose end is a statement as to the patient's condition. In a confusing manner, the final statement is also called "diagnosis". In the process of diagnosis, examining physicians listen to patients and ask them questions. The answers to these are called symptoms. They also observe the patients and their observation reports are called signs. Like the term "diagnosis", the term "symptom" is also confusing: sometimes it is used to mean symptom, and more often, symptom or sign. Before diagnosis is completed the patient's symptoms are endorsed as they stand or in modification, or else the are omitted (usually by the examining physician); this way all the patients' processed symptoms may be viewed as signs, since they are now the examining physician's observations. (In the present discussion only the word "symptom" will be used, except when its various senses should be distinguished, and then it will be made clear which one is intended.)

The diagnostic process involves tentative attempts to group signs and symptoms together into descriptions of familiar, more or less standard clusters called "syndromes", and the final syndrome constitutes the diagnosis, namely, the final diagnostic statement. The diagnostic tools are the latest medical techniques for eliciting significant signs and the latest medical theory that enables one to decide what is a significant sign and that helps correlate symptoms with syndrome, syndrome with disease, and disease with its cause. The correlation of disease to cause is called "etiology". Applying all this, diagnosticians decide from what disease and/or etiology patient suffers; upon such a decision a diagnosis is completed and a recommendation concerning treatment follows. Traditionally, the break between the decision regarding the nature of the ailment and the recommendation concerning treatment is clearly delineated. The recommendation is called "prescription" or "recipe" and belongs to treatment. Yet in between come many other factors, especially course, the description of the natural course of the disease, and prognosis, the course of the disease given the compliance by the prescription, but also economic and other considerations that in the light of course and prognosis help choose between available prescriptions (including the null prescription, the prescription to do nothing about the disease). For example, it is not clear what will happen to patients if they follow the null treatment and are invited to return to the clinic if the ailment persists;   or if they follow one of the available prescriptions on the understanding that if treatment is not satisfactory another mode will be tried instead.   Diagnosis and treatment intertwine within the clinical context.

Diagnosis, the diagnostic process, is supposedly confined to initial clinical diagnostic encounters.   In more elaborate cases the diagnostic process includes brief laboratory examinations of samples and specimens.   When examination becomes even more elaborate and the patient is hospitalized for a few days, the process ceases to be viewed as merely diagnostic even if it is, since, in this case the diagnostic procedure may become complicated and incur a high price.   Costly diagnosis is traditionally considered within the context of treatment. Examination, when distinct from treatment, is in a clear sense all diagnostic, yet tradition puts a limit to encounters considered purely diagnostic.   Although monitoring a patient under treatment is diagnostic, it is not seen as a part of diagnosis.   It is rather viewed as an integral part of the whole therapeutic process.   What is the line between a brief and a prolonged encounter for the purpose of examination?   This question is traditionally left open, and for a good reason: there is no urgent need to draw this line.   In any usual study of a computer diagnostic service, this is particularly the case. Technically, the object of such a study is the brief initial diagnostic encounter.   Therefore the problem of the limit of initial (or any other) diagnostic encounters is usually entirely ignored, and quite rightly so.

This limitation may be considered an idealization that fits only some purposes of diagnosis (and of the use of computers in medical diagnosis).   A more general analysis takes diagnosis and treatment as intertwined and thus each provides the context for the other.   There is no need to differentiate between brief and prolonged examinations or between examination and monitoring.   Taking treatment as the context for diagnosis is important in two respects.   One concerns the study of diagnosis in general, which is the objective of this book, and the other concerns technical aspects.   As to our objective, our main goal is to improve treatment rather than diagnosis; diagnosis and thus also diagnostics are mere means to the task of medicine. As to the technical aspect, including the matter of the monitoring of treatment as a part of diagnosis, it is made possible by the very general character of our objective, namely, the emphasis we lay on the formal nature of the computer part of the computer-assisted diagnostic service. This facilitates exploring the use of both the formal aspect of the computer service and the ability of its user to bridge gaps in that service.

The systems-theoretic techniques engender this kind of generality. Whenever the context of any activity is too limiting for a given purpose, the eye is naturally cast on the para-text in search of an item from it to transfer into the context, at least experimentally. Experiments of this kind are guided by inherently vague rules and, therefore, inherently they do not yield to computer programming. Yet they are eminently open to assistance of computer programs--since these can easily parade and then eliminate many candidates for altered contexts. In diagnostics the latitude just offered is not merely one permitted and assisted by the computer. It is a latitude scarcely available without it. Considerations of this situation render the tremendous advantage of using the computer readily apparent.

Consider the following situation: it is well-known that both a syndrome and the correlation of that syndrome to a disease is a matter of mere probability. Details of information are often missing, or are linked together in an unorthodox manner. When the medical textbook declares that some signs and symptoms of a given illness appear together with a given sign--e.g., a high temperature--and a patient exhibits all of them, yet without the expected given sign--e.g., the patient has a normal or low temperature--it is then a matter of probability whether or not the patient has the illness which the medical textbook describes. But even when the given signs and symptoms are exactly those described in the medical textbook as one syndrome, the syndrome is not always correlated with one illness only. In the face of uncertainty, likely candidates for the illness would be arranged, given the syndrome, in the order of their likelihood; and then a search is under way for a sign or a symptom that will tip the scale in favor of one of these. For example, a syndrome may be caused by a virus or a bacteria. A microscope or a ready-made chemical preparation will then be used, and puss in a specific organ or sample will be observed, in order to differentiate the two. This process is traditionally labelled "differential diagnosis". But, differential diagnosis may not be an available process because there may not be an easy method to decide between two options.

The use of the computer is now rather obvious. The computer may offer a list of options too large to handle otherwise, and probability estimates for each disease, given the syndrome which a given patient suffers from. Alternatively, the computer's meta-text may list para-texts from which to choose a context, and then offer probabilities within the chosen contexts. The process of diagnosis usually involves series of statistical considerations. As was mentioned before, diagnosticians may not know the extent to which the factual information at their disposal is reliable. The excessive use of statistics leaves too little room for the bare facts to carry much weight. Yet, without any statistical considerations, the cost of every diagnosis increases drastically in time, money and suffering. Even when differential diagnostic techniques are known, cost is variable and may be prohibitive, depending on the technique chosen for application and the circumstance. Here diagnosis and treatment intertwine within the epidemiological context.

Human ability to perform adequately under uncertainty has been investigated and found scant. The study has evolved over the past four decades and it constitutes a respectable science today, the science of psychological heuristic. Psychological heuristic examines the way people perform under conditions of uncertainty, how adequate their performance is, and how much of that performance is given to precise formulation which may help people under conditions of uncertainty. Here is an example of an eminently inadequate performance taken as it stands from D. Kahneman and A. Tversky.

A well known logical truth is that whatever qualities are considered, the probability of having one of them is never smaller than having both. Kahneman and Tversky elicited from a large portion of samples of subjects probability estimates which flatly contradict this logical truth. What they did was simply construe counter-intuitive examples. When an individual is described who is known to have a low likelihood of possessing one quality and a high likelihood of possessing another, unrelated quality, then regularly a large portion of subjects studied--including subjects familiar with the calculus of probability--declared it less likely that the person has the one quality than both. It may seem commonsense that subjects might compensate for the distasteful information that the person described has the first quality by declaring that person likelier to have both the distasteful and the pleasant qualities than the distasteful one alone. Kahneman and Tversky explicitly claim their findings to be neutral to a psychological reading of this sort.

Kahneman and Tversky have found that at times physicians fairly unanimously judge the probability of a person having certain two characteristics higher than the probability that the same person has one of them. When told that this contradicts the laws of probability some of them said, it may be true in theory but it is false in practice. However this experiment is judged, it clearly is thought provoking.

The experiment indicates that the paucity of evidence becomes increasingly problematic as one adds to extant evidence some irrelevant items. (Hence the trouble with diagnostic ballast: it is logically irrelevant and practically harmful.) Experiments like the ones conducted by Kahneman and Tversky are of importance, particularly in the domain of medical diagnosis: medical diagnosis is often statistical; diagnosticians often consider intuitively factual information and analyses of their probable causes, and they may easily employ faulty intuitions. A very popular computer service may beneficially and cheaply replace intuitive assessments of probable causes with precise calculations.

We do not know why people misuse redundant information. But consider again the example just cited of the probability estimate of a given individual's unlikely quality and likely quality. Possibly the subjects asked confuse the two questions--concerning the individual described as having the one quality once with and once without the other quality--with the two questions concerning the same individual having either the one quality or the other but not both. To spell things out, consider the following possible cases:

(a) Tom is tall;
(b) Tom is tall and thin;
(c) Tom is tall and not thin;

Theorem: The probability of (a) equals the sum of the
             probabilities of (b) and of (c)

Question 1. Is (b) more or equally or less probable than (a)?
Question 2. Is (b) more or equally or less probable than (c)?

When people are told a story to illustrate case (b) they may
be misled to think that the they are expected to answer
Question 2., whereas they are expected to answer Question 1.
The psychological overtones of the story are designed to
provoke this misunderstanding, as the stories conflict with folk
wisdom. Kahneman and Tversky may very well have noted
this point: they try to help their subjects correct themselves;
but it is not easy to accept corrections. Why? Is the case
Kahneman and Tversky describe a systematic error or is it an
expression of psychological unease? Our own view is as follows.
    Different studies suggest that the following grave error is
rather common: use a set of data to surmise a probable
hypothesis, add the hypothesis to the data as a given datum,
and use the expanded list to estimate further probabilities.
Different studies indicate that even scientific researchers at
times behave in this way, though it is known that this way
any desired probability estimate is achievable. Moreover, this
process may be accomplished in different sequences, usually
leading to very significantly different estimates. This occurs
too often in research laboratories and in medical circles of the
best quality. We suggest that the subjects of the Kahneman
and Tversky experiments did just this. We regret that they
do not deviate enough from inductivism to permit themselves
to present alternative hypotheses to explain their practically
important findings in order to put them to crucial tests.

3. THE TASK

What makes a diagnostic judgment reliable? When is a
clinician trustworthy? Expert clinicians repeatedly make quite
complicated choices of probabilities, not always correctly. It is

well known that the experts do not bother to calculate probabilities precisely. Perhaps medical intuition helps them make good decisions without calculating probabilities. Designers of computer programs for medical diagnosis often aim at a level of performance no higher than that of the specialist. Can the computer help here and improve diagnostic performance further by starting a diagnostic custom of relying on precise statistical computations? The intuition of diagnosticians concerning probabilities may be improved upon significantly by the use of computers at almost no cost, yet this may have no impact on the end result. It is not at all clear, however, that improving probability estimates will improve diagnostic performance. In other cases the service of computers might even be harmful, perhaps by destroying all intuition, possibly intuition that can rectify errors that computers cannot detect. As the computer is only one mode of tightening the rules, all these questions go to the heart of any study designed to examine the advisability of tightening rules.

Social considerations may be likewise problematic. For example, the implementation of a computer service is always done at the expense of other aspects of medical service, and these may be more useful. Excessive reliance on computers may cause or at least encourage new kinds of neglect harmful to patients; by some remote reason and by some irrelevant error this defect may become socially acceptable. We have examples for this from the use of life-sustaining artifacts which extended the lives of brain-dead patients and of hopeless patients: this use has led to the attempt to revise socially accepted attitudes towards them. Hence the relevance of social and moral considerations to the study of the implementation in medicine of any new set of rules and of techniques, including those that go with computer services.

There are, however, benign aspects of the use of computers in diagnosis that no-one has questioned. The computer may assist patients in some self-diagnosis; it may assist diagnosticians by acting as a reminder, as a swift calculator, and as a safeguard against certain common errors--at times also as a quick source of information: no one knows all the currently established rules of diagnosis. Indeed no one knows them so well as to be able to feed them to computers. But some kinds of failing and weakness are known which

computer services have proved able to help rectify. As long as
the computer does not impair judgment or raise costs more
than the cost of some more vital service, it poses no particular
obstacles; our initial dilemmas are thus resolved, at least until
some obstacles are discovered.

Existing computer-assisted diagnostic services intrude
significantly into the traditional diagnostic encounter. Both
patient and diagnostician are the customers of the service, and
all too often there is an inevitable conflict of interest between
them. It is thus important to determine who is the customer
and beneficiary of any item of any computer diagnostic service.
In the present book we dwell at length and repeatedly on this
question. As long as the rules of diagnosis in general and of
the use of computers in particular will remain much more
readily available to physicians than to patients, the service
which computers might render will remain underutilized.
Diagnosis would be greatly improved were it equally available
to both physician and patient. Medical services at large would
thereby improve as well.

This leads to questions of economic feasibility and of
moral propriety of computer service in diagnosis. Economic
feasibility of the computer is fortunately becoming increasingly
unproblematic, since increasingly sophisticated home and office
computers are now widely available and since they may offset
the skyrocketing costs of medical treatment. They offset these
costs by improving both diagnosis and the monitoring of
treatment. But they may generate other costs, some of which
are related to early diagnosis and preventive treatment. These
raise the question of moral propriety which becomes ever more
difficult. The difficulty is general, but it is experienced most
conspicuously where medicine touches the edge of life. With
increasingly sophisticated achievements of medical research,
deciding cases of life and death increasingly fall in the domain
of ethics. The question central to our discussion is, How far
can we entrust decisions of life and death to computers?

To conclude, when the diagnostic process is taken as
traditionally conceived, then the efficacy of any change is hard
to assess--including the efficacy of using computers, or even of
the increased formality of the control of the established rules of
the process. Such matters cannot be properly judged as long
as their side-effects are unknown, as long as there is no

assessment of costs, moral propriety, and benefits, and thus the extent to which it is advisable to increase the formal character of the process. Each of these aspects can enrich the context of the deliberation into the way to improve diagnostic practice. This discussion requires a more precise and detailed framework.

Chapter 3

THE COMPUTER REVOLUTION

IN MEDICINE

1. A DEPARTURE FROM THE PAST

The computer is only a few decades old, and its extensive commercial use is barely two decades old. Its use in medicine is far more limited than in trade and industry. To date, software programs available for any medical use are a mere handful. In every other branch of the economy of comparable size, the market is flooded with many alternatives. The use of computer-based machinery in medicine, such as the highly publicized computerized axial tomography (CAT) scanner, is much less extensive than in other industries. Only recently, for example, the robotic hand, originally designed for the automobile industry, was adapted for brain-tumor removal. Most computer use in medicine is limited to a place on the clinic receptionist's desk and not in the diagnostician's clinic. Yet the computer has effected a quiet revolution in medicine. Its impact is exciting and extremely beneficial yet one cannot responsibly overlook it as a potential source of great worries.

A review of the literature on the use of computers in daily clinical diagnostic encounters gives rise to concern and puzzlement. There have been a few teams of Artificial Intelligence researchers aided by medical practitioners and researchers who have seriously attempted to develop pioneering systems of software in the service of diagnosis. These developments convey an unrealistic, excessively optimistic outlook on the whole. Reports on them are presented as research progress reports, yet they rarely supply any evidence of critical examination of reported results. This literature has engendered confusion. Some of the confusion is obviously quite insignificant, but some is dangerous, in particular, the confusion of two senses of the term "decision". Within computer technology, this term denotes a decision by a machine concerning provability, which is set according to some predetermined rules, including, at times, some rules-of-thumb which may lead to some hilarious proposals. (This is known as the "brittleness" of programs which "crack" when applied beyond certain, unspecified limits.) In contrast, the term "decision" in decision theory denotes a choice made by an individual in accord with some general policy: the cost and benefit of each option if it is the right choice, and the cost and the benefit of the same option otherwise are calculated and help translate the general policy into specific decisions. (Most of the interesting problems of decision theory, incidentally, are unsolved, perhaps unsolvable.) The term "decision" may give the impression that we can feed a machine with decision theory, personal policies and data, and then allow the machine to decide for us. Many believed this proposal was ideal. We consider it the major danger which accompanies the computer revolution: the relegation of responsibilities to machines (even to ranges well beyond the limits of applicability of the brittle programs which run them).

The discussions concerning the introduction of computers into medicine enhance the confusions concerning this introduction. The computer seems to explode certain current myths which are remnants of the once significant and now superseded inductive philosophy of science. Many traditional thinkers are trying hard to resuscitate that philosophy. Here are some of the facts which they have to handle.

Empirical evidence repeatedly shows that excess information confuses diagnosticians and that medical histories regularly contain much ballast and many inaccuracies. Traditionally it was deemed very unrealistic to try and alter the habits of diagnosticians, except through the invention of new and powerful diagnostic means. This fact is now alterable, especially with the introduction of the computer into medical education. Yet the question remains, Why is it so hard to get rid of harmful excess information, misinformation and even systematic dysinformation?

The answer lies in the popularity of the view--or rather the myth--that scientific theory rests on data. Ever since the rise of modern science over three centuries ago, it was taken for granted that science begins with the indiscriminate collection of facts. Philosophers have at times called this inductivism, or the belief in inductive logic or in the inductive method, or pure or extreme empiricism. Empiricism in one sense or another is endorsed by all except those who follow intuition unchecked by any empirical information whatsoever; not so extreme empiricism, which is the demand to endorse only purely empirical information, that is to say, the demand to reject any information found with the aid of some conjecture. This is a myth based on the important fact that unless a theory is known to be true, its use leads to bias--to some erroneous sifting and ordering of data and even to the erroneous observation and description. This bias may easily be self-perpetuating since everyone wishes to have one's beliefs in one's theories confirmed, and the bias generates its own confirmations. According to the myth, this is a trap which cannot be avoided. Hence, one must begin with data untainted by theory.

According to the inductivist myth, a true theory emerges in stages out of the raw data. The theory is then verified at each stage to insure the avoidance of bias. Thus, says the myth, research must be slow but sure and its product must grow naturally from observation. In the case of medicine the inductivist myth prescribes the course of theory from anatomy to physiology to pathology to nosology; it prescribes the course of practice from symptoms to the diagnosis of syndromes, of disease entities and of etiologies. Practice, thus viewed, is the application of verified scientific knowledge.

The situation is very interesting. Physicians often reiterate the myth of induction and from it they conclude that diagnosis precedes treatment. They are therefore often surprised to find that diagnosis often interacts with therapy. At other times they take it for granted that diagnosis and therapy interact: they observe this as a regular feature of medical practice. How is it, then, that they know, and are yet surprised by the same feature of their practice? This phenomenon has been labelled by Arthur Koestler "controlled schizophrenia". How is controlled schizophrenia possible? How can one both know and not know the same item?

Whatever knowledge is, it doubtlessly brings forth systematically the disposition (1) to act according to it and (2) to answer correctly some of the questions to which it is relevant. There is *bona fide* tacit knowledge, acted upon but not given to articulation; and there is *bona fide* restrained knowledge, recognized as true but not acted upon. Tacit knowledge is exhibited by experts in many fields; restrained knowledge may be restrained by the law, especially in medicine, by neurotic internal regression, such as in hysteria, and by technical limitations, e.g., of scientists on a desert island. These situations are not necessarily problematic, yet they become problematic when conduct and thought are incompatible. One may knowingly act the wrong way (for any reason). Alternatively one may behave the wrong way unknowingly--for example, one may unknowingly behave in a kind of controlled schizophrenia, for example by holding (unnoticeably) in different circumstances contradictory views. In that case one may answer the same question differently in different circumstances, say in the clinic and in the laboratory, and not see this. In the laboratory it was the overcoming of such controlled schizophrenia that led to some discoveries (such as that of the laser). Such developments are surprising. Almost any practicing physician would be greatly and very beneficially surprised to observe the prevalence of pure empiricism coupled with hypothetico-deductivism and to realize that the two are incompatible. How, then, is controlled schizophrenia possible?

The symptom of controlled schizophrenia, of a controlled yet unnoticed inconsistency, is that of repeated surprise. When items from one set are accidentally perceived in the other set, a sense of surprise is bound to develop. It may lead to a revision of the theory or of its application (such as the theory of induced emission in the case of the laser). But this is the exception and the break-through. Otherwise, things return to normal, which includes the return to the disposition to be surprised when items from one set are present in the other. This disposition is not within our awareness, of course, yet it is detectable in the sense of discomfort displayed at the very possibility of transferring an item from one set to another. We will now venture to present diagnostics as such an example.

Diagnosis rooted in etiology offers sharper results than otherwise. Diagnosis is routinely monitored in follow-up observations of treatments, so that specific applications of diagnostic knowledge to individual cases are checked. If etiology is certain, this checking would be redundant. Yet clinical practice often requires altering etiological hypotheses when predicted goals are not attained. Why is the etiological base of diagnosis reassuring? When is such reassurance justified and when does it beg the question?

Inductivism or extreme empiricism allegedly guarantees the avoidance of all error. It thus makes monitoring for diagnostic purposes superfluous. It also judges all error as the result of jumping too quickly to conclusions and thus as culpable. Every practicing clinician knows that the avoidance of all error is impossible, and that some errors are more harmful than others and that some harmful errors result from responsible actions while other harmful errors are due to intolerable negligence. Yet they seldom conclude from this that extreme empiricism is an erroneous philosophy, that the use of hypotheses should be tempered by criticism, and that criticism helps avoid irresponsible error, though not all. Why, then, is it so hard to give up some erroneous view? Why is controlled schizophrenia preferred?

The answer is this: the views held in controlled schizophrenia have different functions. When physicians' status is questioned on general grounds, or when their general attitude is questioned (such as their support of science, and such as their paternalism), they meet opposition with pure empiricism. When their responsibility in a specific case is questioned, whether in conference or in court, their answer has to be specific; the physician under scrutiny can show that the error was due to a normal shortcoming--of time or budget--or due to a practice common in the profession at the time it was made. As long as this situation is admitted, controlled schizophrenia will persist.

Controlled schizophrenia becomes unacceptable when a view in one area impinges on a view in the other. In the present case, it is rather easy to use inductivism as a defense of the profession as faultless, and hypothetico-deductivism as an exemption of certain errors as not irresponsible. But when inductivism imposes more errors in diagnosis than necessary, it is time to make the required adjustment and let go of the harmful myth of induction.

Tension grew between inductivism or extreme empiricism, between the demand that data should not be jettisoned, and the empirical observation that excess data may get in the way. This tension was in part eased by the extensive use of computers for storing useless information. Attempts were made repeatedly to correlate data, in accord with the myth that random data lead to theory. Such attempts failed repeatedly. Etiology was then used; knowledge bases were thus introduced. This move was part of the earliest research (only two or three decades old), in which the possibility was considered of using the computer as a diagnostic tool. The move began a revolution, away from the inductive method and in the direction of the hypothetico-deductive method--including at times the generation of causal, i.e., etiological, hypotheses plus their empirical tests. This, naturally, introduced statistics into the study.

Statistical theory deals with distributions: How are given populations distributed within other populations? (How are the two genders distributed in the American adult population?) It deals with random samples and with sampling methods. (The sampling method picks up samples, e.g., those whose names are drawn by roulette, on the assumption that the distribution of the genders is the same in America as in the random sample.) Applied statistics is the use of statistical theory in practice.

In many instances sampling is sidestepped and statistical methods are applied to existing sets of data. Since these data are not random samples, the results are misleading. Statistical theory has nothing to say concerning non-random collection of data. Random data are amenable to repeatable experiment: the claim that a sample is random implies that by repeating the same sampling method a sample with a similar distribution will be found. (Here the whole sample is considered one fact, and until the observation is repeated it is not scientific!) Genuine statistics treats hypotheses concerning distributions within given ensembles: alternative hypotheses may be tested by reference to facts; the facts that can serve as tests are the distributions within observed representative samples. What are representative samples? What do they represent? They represent the ensembles: they supposedly have the same distributions. This supposition may be questioned. Those who question a representative sample may repeat the sampling. Those who question the sampling methods may vary them. A statistical fact is scientific if and only if it is repeatable; it can then serve as a test for a statistical hypothesis. As opposed to this hypothetico-deductive view, the believers in traditional inductivism claim that theory emerges from facts: collect data randomly, and this will ensure the data being representative: statistical calculation can be performed on them. This is how statistics with no random sampling gain credibility: through the traditional inductivist view of scientific method.

The hypothetico-deductivist statisticians have offered some severe criticism of the inductivist view of scientific method, and its defenders have retorted thus: they use the probability measure which applies to all the available facts, not the distribution which applies only to random samples. This probability is the degree to which one is ready or should properly be ready (these are two variants of inductivism) to assent to given hypotheses considering all available data. (The permission to omit data is the permission to introduce bias, of course, by omitting data unfavorable to the preferred hypothesis.) This is known as subjective probability, which is distinct from hypothetico-deductive, objective statistics.

The subjectivist way of viewing the statistical formula, subjectivism, is refuted by a simple repeatable fact: random samples are used in practice and in research; random collections of data are not: they comprise sheer ballast. The facts used in statistics are repeatedly the specific observations of hopefully representative samples; facts are sampled in the light of competing hypotheses. This is an empirical fact systematically overlooked by the subjectivists. Their oversight is erroneous by their own lights.

Data collectors who rely on heaps of information, see the computer as a godsend. But the computer cannot transform data from its data banks into random samples; only random samples, data collected as parts of tests, are controlled and thus possibly reliable; other data are not controlled and hence are irretrievably unreliable. Data from data banks cannot be considered random, since their randomness cannot be checked: randomness is checked relative to given hypotheses during the sampling process: the sampling is devised to insure randomness. Hence statistical hypotheses must precede observations: new hypotheses prescribe new randomness tests and so, usually, new sampling and new observations as well. All this is known standard procedure: the use of old data to test hypotheses is only as good as the initial test, and neither agreement nor disagreement between hypothesis and old data count for much more than heuristic clues to researchers. As this is standard procedure, the contrary view is sheer myth.

Though subjectivists generally deny the existence of these objections they try to meet them with diverse hypotheses. These hypotheses are self-defeating: Should they be tested by random samples or by extant data? Has any of them ever been tested in any way at all?

This objection shows that the more subjectivists attempt to handle objections the less practically relevant their subjectivist theory becomes, since it is improper to apply the subjectivist theory prior to tests of its excuses. Thus, subjectivists come with a supreme practical argument: as long as there is no adequate scientific theory guiding conduct in specific circumstances, such conduct must rest on the bulk of extant data. (This is the view adumbrated, for example, in the most popular and authoritative *Diagnostic Statistical Manual III-R of Psychiatry* of the American Psychiatric Association, to be discussed later on.)

The subjectivist's practical advice to use data in lieu of hypotheses is inconsistent with common practice: conduct not based on proper scientific theory is usually based on conventional guidelines. Treatment not based on theory is the grey area where there is room for patients' complaints, often leading to litigation. Law courts endorse any treatment based on generally recognized guidelines; they do not recognize practices based on an honest subjective assessment in the light of unrepeatable data.

One may object to this and claim that things are different now, since the computer can offer us tremendously large data bases. Inductivists are, indeed, impressed by multitudes of data. The fact remains. Law courts, guided by expert witnesses, recognize either the extant convention, or the proper introduction of new treatment based on scientific hypotheses checked by random samples. Subjectivism is not a recognized practice; it is a myth held by individuals who reject it under cross examination.

2. THE CURRENT SITUATION

Advocates of inductivism ascribe the success of computers to their capacity to process great quantities of data. Advocates of hypothetico-deductivism ascribe the computer's success both to its ability to process relevant data and to the usefulness of this service for critical thinking. Which of these two assessments is true? Let us look at some data, beginning with those which seem to support inductivism. As an example let us explore facts which concern diagnosis.

Some of the diagnostic expert-systems software programs which have been developed in order to assist physicians in generating diagnostic and therapeutic hypotheses are inductivist in orientation: their originators are sustained by the hope to secure novel computer-generated hypotheses, supposedly to emerge out of raw data. This is the desire to revive inductivism, as a tool for both discovery and practical application. A team of Artificial Intelligence researchers have developed a software program which supposedly prompts the computer to extract from given data some laws of nature. (The program has excited scholars; it is called "BACON", after Sir Francis Bacon, the inventor of modern inductivism). A similar, diagnostic program invites the computer to develop a new rule of thumb to choose a null hypothesis about the condition of a new patient. In addition, there are computers programs recording what is called "confidence factors": the computer instructs the physician to type 1 to indicate absolute faith (in some hypothesis), to type 0 to indicate absolute disbelief (in it), and to type 0.5 to indicate total indecision; when half-inclined to agree, one may be instructed to type 0.75; and so on *ad lib*. Since this proposal is advocated these days by leading students of clinical theory, we shall take it up later in this book, even though we are rather cavalier about it. Yet we confess we do welcome computer-prompted relevant hypotheses: we recommend that as many hypotheses and schemas for hypotheses be fed into the computer in a manner that will enable its user to order it to suggest hypotheses as efficiently as possible. But hypotheses have to be tested before they may be applied: confidence is no substitute for tests. In reply to this contention it is repeatedly asserted that confidence is but the result of data, or that it is justified when based on data, that justified confidence is a measure of appropriateness of the practical application of the justified hypotheses. But the appropriateness of hypotheses is the permission to apply them, and the permission to apply a hypothesis always rests on the claim that sufficiently many and sufficiently severe prior tests have been performed in accord with convention (so as to eliminate obvious errors). And convention properly demands that new hypotheses be tested in accord with legally prescribed specifications. When there is permission to apply given hypotheses physicians may and often do act on them regardless of any possible lack of confidence, and conversely, confidence is no substitute for legal permission to act, nor for the legal procedure that may secure it. In reply to this contention it may be said that the rules of induction justify the law. But this reply makes the law perfect or it makes the rules of induction imperfect; one way or another, it should be stressed that the law has to be complied with even though it is not perfect and should be improved when its imperfections get in the way in a drastic manner.

The significance of the hypothetico-deductivist assertion that hypotheses precede data and never emerge from them is not only philosophical and psychological. It is of a supreme practical import. It not only suggests that the computer can only propose hypotheses fed into it, and never invent ones; it also suggests that the data are tempered with and therefore should not be trusted but rather used to express suspicions concerning the shortcomings of hypotheses. The tampering with data is most conspicuous when fed into computers. To be useful--for practice or for research--data must be standardized. Standardization can be made arbitrarily for the convenience of the users of information (so as to offer them simple access), whether the users are administrators, practitioners, or researchers. Alternatively, data can be standardized in accord with simple theory. When standardized in accord with a theory the data are biased. (This is the traditional and quite satisfactory definition of theoretical bias.) When biased data are used in the field, the result is less likely to be beneficial than harmful unless the theory is wisely chosen. In research some data must be used to test some theory--in which case the bias is set so that information may be inconsistent with the theory to be tested. When the information is sought so as to find a contradiction between theory and data, and this search fails, then the theory is corroborated and when the corroboration is up to the accepted legal norm, then it may be used for diverse purposes, including standardization. Yet the standardization of data for practical purposes has to be corroborated by accepted standards of corroboration, of course: standardization at large is one aspect of applied theory.

Standardization is an institution, social and legal. In the United States, the legal guardian of the process of testing any as yet unauthorized innovation in matters medical, is the Food and Drug Administration. In cases of controversy, things may be aired in law courts, which use expert witnesses. Inductivism presents science as personal rational belief and technology as rational action which is the application of rational belief. These leave no room for the law; they are thus inconsistent with all the data and with all the practices of standardization, social and legal, in medicine and elsewhere.

The earliest service computers provided in clinical diagnostic encounters was their service as memory banks; they compiled data on patients, including their daily progress, medical histories and similar data. This initial service helped to underscore the need to standardize medical information on a national scale. Standardization is either capricious and subjective or based on some theory; it is either subjectively or theoretically biased. The theoretical bias may be as invisible as water to a fish, especially when tacitly and universally accepted and when introduced early in (medical) training.

In general, the very introduction of computers as tools of instruction to medical schools whose students are increasingly computer literate, has had a tremendous impact on medical thinking. These educational computer programs encourage students, from the very start of the diagnostic process, to think boldly, diagnose etiologically, make alternative hypotheses, and exercise their critical ability towards them, and to develop tentative broad outlooks. This is facilitated by software which offers sets of conflicting approaches and hypotheses that enable practitioners to make up their own minds.

This is the crux of the matter--the question of diagnostic decisions and of responsibility for them. The first view of the use of the computer in diagnosis came from the students of Artificial Intelligence. Many of them were adherents of the myth of extreme empiricism according to which scientific theories evolve from facts without the use of the imagination, and so they imagined that computers can be developed which will adequately emulate science and technology--including all the functions now performed by physicians. This led to polarization: enthusiastic and euphoric endorsement as opposed to the fear of and hostility to the vision of computerized diagnosis. Then the Artificial Intelligence community has attenuated its claims: computers can assist physicians, not replace them. This has helped replace the fear and hostility displayed by the medical community by a courteous endorsement of the computer. This endorsement is commendable but insufficient: some of the hostility is justifiable to some extent and this should be taken care of. The computer complex has planted dangerous myths, including the following: computers never err; the proper use of the computer frees the diagnostician from responsibility for diagnostic error; resistability for all computer-generated error lies at the door of the designer and programmer whose work had led to that error. These myths have already been put to harmful use.

The chief harm which these myths cause is in their encouragement of irresponsibility. The irresponsible use of the computer, however, is greatly compensated by its responsible use. An excellent responsible use of the computer involves exploiting its service as an excellent alarm-bell; this is repeatedly exhibited by its use in this capacity in a wide variety of medical monitors. It may likewise easily help weed out some of the worst diagnostic misjudgment. (The diagnostic expert-systems program serving a standard illustration for this is the one which is reportedly able to help improve the detection of digitalis toxicity.) Nevertheless, our concern here is to prevent the standard misuse of the computer as an excuse (including the excuse that the monitor's alarm has failed). In contrast, the misuse of any instrument often involves individual misjudgments and poor overall performance. Researchers tend to exaggerate the success rate of the use of computers in diagnosis and to overlook the possibility of over-trusting it. What is missing here is effective monitoring of the use of the computer and a good alarm-bell, a warning mechanism for the early detection of any failure of the system.

Computers in the field of medical diagnosis may be all too susceptible to abuse, particularly because of two large powerful interests--of the medical profession and the computer industry. In the case of the implementation of computer technology, as in that of new drugs, its benefit and risk should be assessed and monitored. Standards of permissible error, in terms of extent and of frequency, have to be legally decided. Controls over them should be instituted, and research aimed at their improvement should be secured and regularly reapplied so as to raise the standards of performance as high as reasonably possible. Researchers need the freedom to apply controversial ideas, vague ideas, and even sheer fantasy; these are parts of scientific controversy, and they are essential in research; but before they enter into practice, especially on a large scale, they have to pass tests, and by standards that have to be assessed and reassessed. Such standards are yet to be determined for the use of computers in practical matters--diagnosis included. Quite arguably, these standards should be specified by an independent public body and regulated in the same way as in the case of new drugs.

The computer, then, is bursting into the medical world with hosts of problems which are scarcely attended to, largely due to the prevalence of the two competing methodologies, inductivism and hypothetico-deductivism. Both planners of diagnostic practices and practicing diagnosticians need a new framework within which to deliver solutions to the problems which emerge in the new situation, to decide the order of their priorities, and to ascribe new responsibilities to old and new institutions.

A full account of these problems requires more careful analyses of the diverse medical and legal practices extant than can be offered here. The following example illustrates this. It is unreasonable to expect from an average physician to be well-versed in all current theories--not even in their own specialties. Most physicians often have only a smattering knowledge of the relevant theory and they make do with a list of prescriptions based on it. Since they are not well-versed in the theory, they do not quite comprehend the rationale of these prescriptions; when in doubt they have to call the expert; when the expert too is in doubt, then it is an occasion to call a conference. Applying this to the case of computerized diagnosis is imperative. And it is rather problematic.

The problem at hand is of the reading or interpretation of computerized data. When a new diagnostic device is introduced into medicine, for example the x-ray shadowgram, the data it generates (the shadowgram images), are usually (not always!) open to fairly intuitive interpretation; hence the problem is not what is seen but how to interpret it. This has changed with the introduction of better devices, for example the very soft x-ray shadowgrams and computerized axial tomography (CAT scan). The situation has become truly dangerous. The intuitive force of the image on the screen is so strong that at times diagnosticians take it for granted that they see a picture the way we see an ordinary photograph or at least an ordinary shadowgram, especially if they are quite oblivious of computer theory, information theory and pattern-recognition theory. This is not the case. The image on the screen is due partly to the input from the patient's body but partly also to the design of the machine and to the technician's fiddling with the dials in an effort to procure an optimal image. Fiddling with the dials intuitively, the technician finds what seems to be an enhanced image. The technician's idea of what is to be enhanced, however, is often vague, since technicians need not know all the different diagnostic problems and all the different enhancements which these require. Expert diagnosticians aware of the interaction of the diagnostic process with the process of image enhancement, may learn to request from technicians to fiddle with the dials so as to try and enhance different aspects at different times. For it is quite possible to enhance different aspects of a picture, to elicit a picture optimal in one respect at the expense of another or *vice versa*. Diagnosticians who are unable to do so, or who are unable to read images anyway, will have to consult experts. In this case experts should be knowledgeable about computers as well as about medicine. Such experts are very rare, and at the present time training in this expertise is scarcely available.

This is but one example, and one quite central to all software diagnostic tools available, though the problem is concealed behind a fancy terminology. These tools are called diagnostic expert-systems software since their designers have built them with the intention of emulating expert diagnosticians. It was assumed that the ordinary specialist performs at a level inferior to that of the expert. It was assumed that with an expert at their elbows, specialists can achieve much better diagnostic results. It was assumed also that the computer can emulate the service of the expert in this very respect, provided that the specialist has a modicum of computer literacy, and is facilitated by a user-friendly program. This is what expert systems are offering, and not only in medicine. This does not look problematic, but it is, and the problem lies in the inability of experts to convey to specialists all they know--or else all specialists would be expert; and this inability hides behind the half-believed myth of induction in the following way.

The hierarchy of expertise is in part a matter of possession of explicit knowledge, which is used in the process usually described as eliminative induction, except that this eliminative induction differs from traditional, strict eliminative induction: tradition guarantees success on the assumption that the initial list of options from which errors are eliminated is complete. A real list of explicit options is always limited and thus its use is always tentative. In practice, a list of the higher expertise includes those of the lower. Yet, without a guarantee, as even the most general list is incomplete, and lists are regularly revisable. The revised list is the generator of revised lower lists for those whose expertise is lower on the hierarchy. The revision is usually offered by research labs after securing legal permit. All these steps would be both unnecessary and impossible were the process strictly eliminative induction, namely one which comes with a guarantee of success. Strictly eliminative induction is inapplicable not only because our knowledge of the possible options is partial and inaccurate; it is also not applicable here because the law takes account of the imperfection of our knowledge and does not permit its application without prior approval.

What would be the best and most explicit diagnostic computer program? Ideally, such a program has to serve as a machine almost eliminative-inductive as well as diagnostic-processing. It should look as if it is the very best expert.

This ideal, however, is somewhat misleading, since it sidesteps the question, When should one consult an expert? The diagnostic expert-systems software offers the same expert strategies and proposals and advice which experts offer to the specializing students possessing limited knowledge of the relevant theories. When a very thoughtful advice for the application of a theory is needed for a very special case, the expertise of the diagnostic specialist is insufficient, and so the expert-systems software is equally insufficient, and consultation with the expert who can apply the theory to the special case at hand is required. To ensure that the potential user is aware of the need for an expert consultant, responsible dissemination of knowledge of the limitations of the software is necessary. It has not yet been determined who are the agents responsible for this activity and what controls should be put on them.

So much for the problem which hides behind the fancy term "expert systems". This problem should not be overlooked where the use of expert systems may concern matters of life and death. Consequently, there is an urgent need for the rapid and detailed development of a broad picture of medical theory and practice, of a critical analysis of the diverse medical institutions and traditions, and of new ideas concerning changes and implementations invited by the integration of the computer in medical practice. All this is partly being effected in the modern world and it is a welcome revolution which the computer is helping the medical profession undergo. Yet even if all these needs were met, it would not suffice.

What more is needed? What more can be done?

The roots of certain unsatisfactory traditional practices may be studied, hypotheses may describe them, traditional practices may be assessed and examined carefully, and the positive and the negative in the incentives which sustain and perpetrate the negative in current practices may thus be discovered. This should lead to some institutional reform which will abolish incentives whose effect is negative and insure the continuation of what is positive in the traditions--so that the reform will have the net result of improvement with minimal loss. This is particularly the case with the desired replacement of traditional inductivism with hypothetico-deductivism in each sub-field of medicine.

The traditional myth of extreme empiricism may serve as a good illustration of this. The myth is potent and highly convincing, and as such it ought to be taught and presented with a critical examination of its strengths and weakness in all medical schools. (The views that no error should be taught in medical school, and no philosophy, are objectionable parts of the myth of induction.) The positive side of the myth is that it discourages the irresponsible application of any new and wild idea prior to its empirical examination and that it offers a justification for any responsible application of ideas. It has severe drawbacks in its discouraging the researcher from developing creative ideas. (Niels Bohr used the colorful word "crazy" when he referred to creative ideas in science: a really good idea, he suggested, must first sound "crazy"; of course, this is an understandable exaggeration in the opposite direction.) It has the drawback of offering a vague blanket justification of all current practices *en bloc*. Instead we should recommend the testing of each practice as best as possible and the improvement upon it when possible, and the application of the improvement only after new tests, in accord with the accepted standards (which also need testing and improvement on occasion, as well as public assurance that occasional tests are honestly undertaken).

To end on a new positive note, the growth of computer literacy in the modern world opens up an abundant new market for computer-assisted self-diagnosis. The computer will, no doubt, easily outdo all older tools of self-diagnosis, mostly popular encyclopedias and popular handbooks of medicine. It will then serve as a tremendous educational tool for raising the general level of medical awareness in the general public. The more diagnostically aware the public will be, the harder it will be to overlook the obvious fact here so repeatedly emphasized: diagnosis is not limited to the initial clinical diagnostic encounter; it is pervasive in medicine.

To conclude this discussion, this book on diagnosis includes repeated reference to the computer because of the tremendous role it is playing in this field, and an ever increasingly central one. As all tools, it can be misapplied. The establishment of new social institutions, including new democratic controls over the use and possible misuse of the computer, may thus be advisable. This is not the place, however, to introduce details concerning the need for the democratic control of technology in general and of every branch of technology in particular, including the field of computer-assisted diagnosis. The general idea, however, is that the raising of the standards of medical diagnosis, and the introduction of the computer software as a diagnostic tool, can be effected together in a natural manner to benefit medicine and society at large.

In closing this discussion, let us return to the application of the most significant point of scientific method to computer technology. The computer may be used in an inductive and in a hypothetico-deductive fashion. When used inductively it stores all sorts of possible data on which statistical calculations can be effected. Such exercises look profound and impressive, yet they are worthless at best. In the hypothetico-deductive mode, simple problems are stated, their solutions are posed as general hypotheses which may be statistical. If the hypostheses are scientific, then they can be tested by repeated observations. In statistics observations are of random samples (with strict tests of randomness): they can be repeated in accord with statistical theory. In particular, one set of problems that have to be solved (by posing hypotheses and testing them) is, what data should be stored in a more or in a less accessible manner and which should be dumped so as to prevent the data bank from clogging on excess data. This sort of problem is studied in the theory of storage and queuing.

Within the medical diagnostic system, the patients are the intended beneficiaries of the proper implementation of the hypothetico-deductive method and the elimination of the inductive method. Yet the immediate beneficiaries might very well be the medical practitioners. The view that diagnosis is confined to initial diagnostic sessions seemingly eliminates controls over diagnosis, which thereby may help conceal some errors. The concealment of error may look personally attractive to anyone prone to error--and everyone is prone to error. Yet publicly this is too costly. Nevertheless this way of handling the fear of error renders it prevalent, and with it the fear of being detected despite evasive maneuvers becomes prevalent as well. The idea of responsible conduct relieves responsible practitioners of such fears while it offers them simple public criteria of what error is irresponsible. This is well-known in cases where significant error is part-and-parcel of certain procedures and there the idea of responsible error is instituted. We shall go into details of this matter when we shall discuss monitoring in some detail. Here we wish to stress that the attitude to error prevalent in cases of misdiagnosis of appendicitis can easily be made most general, and the use of the computer is a golden opportunity for doing so.

3. A PROPOSAL FOR THE FUTURE

Here is a summary of our proposal, to be discussed later on.

1. The diagnostic part of medicine is the weakest link in the chain of the medical services as they are practiced today in the modern world: the improvement of diagnosis is the most promising amelioration of current medical practice.
2. A fully computerized diagnostic system is neither permissible nor feasible in the foreseeable future.
3. A computer-assisted (as distinct from a fully computerized) diagnostic system is conspicuously a three-party system which replaces the current seemingly two-party system. The clinical, social, ethical, legal, and economic implications for the diagnostic process invite careful consideration--especially when these involve computers. Until this is done, extant opposition to the implementation of computer-assisted diagnosis must be viewed in part as rational.
4. A comprehensive analysis of the medical diagnostic system is needed for theoretical purposes as well as for the technological purpose of the rational planning and implementation of diagnostic services in general and of computer-assisted diagnostic services in particular. Implementation can be planned in a manner encouraging growth--of the computer-assisted diagnostic service, of the spread in the geographic area where the service is to be offered, and of the medical specialties to be covered.
5. Any comprehensive analysis and any planned change in the current system of medical diagnosis--computer-assisted or otherwise--ought to touch upon the interface of the medical diagnostic system and the social system at large. Medical diagnosis ought to embrace the study of

individuals in society and of social and political life, lifestyles and the role illness plays within them. This is how the medical profession may responsibly assume a mandate to treat problems that go beyond its traditional scope. This extension of the scope of medical diagnosis beyond what is endorsed by tradition may invoke a new medical lifestyle as well as new uses of computers in medicine and new limits to these uses. It may be very advisable to institute an independent public body for the comprehensive monitoring of medical standards.

6. Any comprehensive analysis and any planned change in the current system of medical diagnosis--computer-assisted or otherwise--ought to maximize the amount of information and the degree of participation open to patients in all stages of the diagnostic process. The option to participate should be regularly available to each individual patient. This way informed consent will become a feature of the diagnostic process. Computerized diagnosis goes against this process; computer-assisted diagnosis may boost it.

7. Personal medical records of all sorts, if significant for diagnostic purposes, should be standardized to some extent. Performance should be monitored and compared to accepted standards, and these, in their turn, should also be open to revision; records should be placed centrally, to be available for any clinical diagnostic encounter at the joint request of diagnostician and patient.

8. Customers should have access to the computerized elements and they should be helped and encouraged to use it. Open access to these elements is an excellent educational tool that will increase the contribution of the service to scientific progress and technological success. Opening access may also improve clinical quality-control and thus encourage the sharing of responsibility by all parties. In addition, this will be free education in the democratic process.

9. The list of possible diagnoses (as well as their possible coincidences) should be arranged with the order of frequency and accessibility--or of the utility of their use (where utility is defined as the product of probability and

expected benefit). Lists should be geared to needs of specific populations. The null diagnosis, the null hypothesis that the patient is well, is very important and problematic and should not be neglected. (It will be discussed below.) Shorter lists should be partially embedded in more detailed lists, in the fashion of dictionaries; this may be called *the onion model of diagnosis*.

10. In addition to practical diagnostic services, expert systems in medicine should be enlisted in the service of studying, challenging and improving the current diagnostic processes throughout the medical system. Diagnostics and computer-assisted diagnostic expert systems intertwine.

11. Existing computer-assisted diagnostic services may be easily linked and thus provide the first stage of the comprehensive system proposed here. The implementation of such services should be regulated by the law to secure their adequacy and propriety.

Our explanation of the above proposal rests on

(1) the peculiar limitations on the diagnostic encounter and
(2) the peculiar role of the computer in the diagnostic process of reasoning, as opposed to its role as a mere diagnostic tool. Since the diagnostic encounter leaves unanswered many questions (for want of time and resources) and since computers can expedite the diagnostic process, it looks as if it is an ideal solution to a vexing technical problem.

The starting point of a proper discussion--of diagnosis or any other matter--has to be current views. There is a principle of diagnosis which is currently prescribed by virtually every medical textbook and in nearly all medical schools; it is too abstract and utopian to be used clinically as is; it should stay in the books, but merely as a regulative principle. The principle is the demand that every diagnostic encounter should lead to a diagnosis which is as thorough and complete as possible. What does this demand amount to in practice? This depends on the reading of the expression "as complete as possible" in the practical context of diagnosis. The principle should offer general guidelines. It does not; worse still, in some cases it is qualified by tacit pre-determined guidelines whereas it should prescribe some, and explicitly so. The tacit guidelines are parts of the medical tradition which override the principle, so that it becomes ineffectual and only generates anxiety.

A few important corollaries follow. When guidelines are articulated, the articulation may be mistaken or insufficient and thus invite criticism or re-wording, perhaps even reform--all of which may be considered positive outcomes. The fear of critical examination is an obstacle on the road to articulation and is a relic from the period in which medicine was scarcely scientific and from the traumas of the transition towards more scientific medical services--a transition which is still under way even in the most advanced industrial countries. The fear of criticism is fairly prevalent and may reflect not only pre-scientific attitudes but also moral sentiments that conflict with respect for the autonomy of patients and of their care takers. These attitudes and sentiments are obstacles to be removed. The removal of the fear of criticism can be facilitated by rendering the task at hand rewarding. The task at hand--articulating guidelines for diagnostic practices--is rewarding, once it is seen as a necessary step in the process of the introduction of computers into the diagnostic process (although even well-formulated general guidelines are usually far from being sufficiently explicit for the needs of computer programmers).

The traditional limitation of the diagnostic process to the initial clinical diagnostic encounter has to be removed. No one fully limits diagnosis to the clinic; its extension is done in an *ad hoc* commonsense manner. Articulating rules for the extension may improve it by the means of democratic critical discussion. Diagnosis will then be defined as a process of deliberation, including deliberation about whether to solicit more input from the laboratory or the hospital (at times the needed input may call for extreme measures such as a surgical intervention); any deliberation that may lead to the prescription concerning treatment is also diagnostic.

It follows that in every stage of treatment, deliberation is partially diagnostic, since it inevitably opens possibilities for revised prescriptions concerning treatment. It also follows that check-ups are diagnostic, even when the treatment they lead to is no treatment at all. More specifically, combining these two corollaries, and applying them to the requirement to articulate rules we may add the requirement to articulate rules regarding the proper moment for the discharge of a patient, or for the hospitalization of (and even operation on) a patient for diagnostic purposes, and so on. The limits of the applicability of such rules also needs to be explicitly articulated and thus improve autonomous decision making that will raise the level of the well-being of patients.

Almost every medical encounter includes a diagnostic component. As most modern medical complexes employ computers, almost no physician-patient interaction in the modern world is free of some involvement with computers. Moreover, almost every modern diagnosis involves computer-assistance of some sort. Yet almost no critical attention has been thus far paid to the question whether the computerization of diagnosis is desirable, not to mention of the desirable degree and style of such computerization. (The question is repeatedly answered by almost every individual involved; yet it is hardly studied scientifically, as one might expect.) Why? In reply one might claim that the computer's role is insignificant as long as its use is marginal (as in simple record keeping). Now many medical establishments still use the computer marginally, but even then the computer's role is significant. Examples will be provided later on of marginal yet significant items of this kind. It is clear, anyway, that a tool may be marginal in one sense and not in another. For example, it may be seldom used, yet save a life every time it is used--or lose a life. If a computer answers this description, then the reply that its use is marginal is unsatisfactory.

Along with the general case--which almost always concerns computers, however marginally--there are cases which involve computers centrally: computer diagnosis *par excellence.* Computer diagnostic services differ from any other diagnostic tool, such as x-ray photography or computerized axial tomography (CAT scan). These instruments are in no way different from the magnifying glass and the stethoscope--they enlarge the diagnostician's vision and thus add to the stock of available diagnostic information. They are not directly involved in the process of reasoning, nor do they form the judgment that the diagnostician may reach.

Taking the input as given for a moment and considering the process of reasoning, the study of diagnosis concerns decisions as to prescriptions of treatments, and it also calls for a new attempt to discuss the principles of diagnosis. These are, almost exclusively, cases of diagnostic expert-systems software programs which simulate the best available medical service in certain restricted areas and help perform the complete job of the diagnostician in an interactive manner. Where such systems do not invite consultation with experts, the practice they advocate may very well be quite dangerous and thus objectionable. The dangers include both the petrification of the service and its bureaucratization (not to mention the use of "brittle" programs beyond their cracking point). Furthermore, a diagnostician may blame the computer and/or the programmer for failure and thus evade personal responsibility.

People who object to computer diagnosis on the ground that Artificial Intelligence systems are objectionable, overlook two facts: first, computer-assisted diagnosis is much less problematic than fully computerized diagnosis; second, computer-assisted diagnosis is not as morally objectionable as is diagnosis by fully computerized systems. But one may object even to computer-assisted diagnosis--on the grounds that the individual patient is unique, that the computer accelerates the bureaucratization of medicine, and that the computer endangers the privacy of the individual patient.

These three objections to the use of computers in diagnosis are powerful; lamentably, examples illustrating them already exist. Hence the objections should be met--yet not by opposing the present tide of computerization. The uniqueness of the individual patient invites democratic safeguards against the petrification of computer techniques and against the tendency of physicians to relinquish their own responsibility to machines. The process of bureaucratization will be controlled not by proscribing the computer, but by organizing the computer services and setting them up so that they can form a national and international network while encouraging competition within it. Such networks can be extremely efficient, but they require special protocols to safeguard the individual and individual privacy. This holds for government, commerce, industry, and medicine. The problem permeates our modern life and cannot be evaded. One way or another, modern society has to take the bull by its horns.

Computers can be used as a means of protecting individual patients against maltreatment and abuse. Computer-assisted diagnostic services can help here--provided they incorporate a comprehensive demand for informed consent, not only for treatment, or for specific diagnostic procedures, but for all diagnosis. The fiction that diagnosis is confined to some initial clinical diagnostic encounter is the excuse for not regularly eliciting informed consent and for not allowing patients to participate as actively in their own diagnosis as they may wish. Once the on-going informed consent in diagnosis is strictly enforced, patients will be encouraged to use the computer-assisted diagnostic services on their own, both for general information and for monitoring their own care.

Computers can be used as means of protecting individual physicians against maltreatment and abuse. Computer-assisted diagnostic services can help here--provided comprehensive informed consent is required. Physicians, especially apprentices, can be informed of their rights and duties and of the specific conditions within which they are invited to participate actively in diagnosis and in treatment of patients within a responsibly organized institutional arrangement. Instead of the burden of the full but hardly usable diagnostic manual, of the full list of all known diagnoses, students can learn to use expert-systems diagnostic lists, where the lists are tentative, and ordered according to revisable utility levels as explained above. Users can then be invited to suggest changes. No doubt, since the diagnostic process itself is not fully formal, these lists cannot be the whole story. The same holds for the manual, though the lists are much more usable than the manual. This might reduce the proverbial pressure on apprentices in emergency-rooms and on call; it might also alleviate some of their understandable hostility to activity under pressure.

Computer-assisted diagnostic services should include sub-systems that might be linked into other networks, both across medical specialties and across towns, regions, and countries. The network should include general information for the lay public, such as glossaries and simple patterns of diseases, simple calculations of compound probabilities that could prevent the errors involved in intuitive assessments of probabilities, and some epidemiological information. It should also store the cumulative records of individual patients under limited access protocol. Planned this way, the diagnostic system would readily assist all its customers with maximal information and encouragement and help in their participation in the diagnostic as well as the therapeutic process--and at minimal cost.

Proposals along these lines may explain some of the anticipated resistance that this book might provoke. A narrow view of the short-term self-interest of the medical profession regards these interests and the public interest as clashing. As in all other cases of clashes between the application and development of new technologies and some special short-term class interests, the public deserves to be informed about matters.

The question with which this section began is left open: How thorough should each diagnostic procedure be? There are no hard and fast guidelines. The general theory for such guidelines is the theory of cost effectiveness. This theory makes use of the whole field of systems analysis, including decision theory, so that its considerations may be integrated into part of the computer-assistance program that should in turn be integrated into the network of public medical services. A country in which such an elaborate and powerful system is available, must establish some kind of monitor to guide and control the use of computers on the national level. This seems to be called for on other grounds as well, and it should hold good for all sectors--government, commerce, industry, and, no less, medicine.

Chapter 4

ETHICS OF DIAGNOSTIC SYSTEMS

1. INTRODUCTION

The process of systems analysis always depends on finding out the beneficiaries of the system, their problems, and their expectations. The present chapter therefore concerns the following questions: Who are the beneficiaries of the medical diagnostic system and what are their rights, specific problems and expectations? This gives rise to the preliminary question: How does one find the beneficiaries to a given system?

The customers of a system are defined as its intended beneficiaries, the ones whom the system primarily comes to serve. Here a difficulty arises, as most systems have several beneficiaries, and attempting to serve them all may raise a conflict of interests. What rule, then, determines who are the *primary* intended beneficiaries of a system?

The simplest way to find out the primary intended beneficiary is to choose the one whose problem the system is primarily designed to solve. Thus, finding out the beneficiaries is only another way of expressing the end goals of the system, and finding out the primary beneficiaries is expressing the system's primary end. But the primary aim of the system can be different in different contexts, each of which can be determined as primary. The diagnostic system includes many sub-systems--the diagnostic encounter, the conceptual framework used, the research sub-system at large, the medical system at large, national health at large, the practitioners, and the clinic. Any of these may be taken as the conditions or the framework for the others. There is no natural order but the one which is determined relative to the problem at hand. The problem is thus restricted to the beneficiaries of the system, which explains why it is important to find out the primary beneficiaries. The primary beneficiaries determine the context of the diagnostic encounter and *vice versa*. The issue is inherently open.

In the present book we take the approach that only individuals are responsible. They, then, are the responsible agents for whatever happens in the realm of their activities-- despite the fact that these activities may relate to public tasks. More specifically, only individuals can act--either as receivers of any service offered to the public at large, or as representatives of the public interest. They are the pivot of any given system in that they are the ultimate judges of any public activity. The guidelines for finding the customer for a given system, then, precede any attempt to analyze that system.

The present chapter consists of three sections that focus on the consumer of the diagnostic system and thus prepares the way for the systems analysis of the system of medical diagnosis to be offered in the following chapters. It explicates the moral principles that guide our approach to individuals--we call them the modified individualist moral principles--and their contribution to the rendering of any medical service that is construed as a system indifferent to morality, to the actual moral practice of medical diagnosis and treatment within its social context, and to their special interplay with medical research. To simplify matters, individual customers will be considered first, and then the corporate customer. We will consider the general public or society at large as a corporate customer, as the consumer of the diagnostic services and these services as largely the products of biomedical scientific research.

2. INDIVIDUALIST ETHICS

It is easy to advocate conformity to one's conscience and it is easy to advocate conformity to the law. A difficulty arises the moment one realizes that private and public rules may be in conflict, that one's conscience may command one to act in a manner unacceptable to society; and here it matters little whether the conflict is with customs, with the rules of the institutions to which one is bound (be it the medical association or the hospital clinic) or with the law of the land. Traditionally collectivist ethics demands that loyalty to one's group (collective) stand above private conscience; classical individualist ethics demands the opposite. Modern individualist ethics differs from both. It is still individualist, but it rests on the understanding that acting in total disregard for custom, institutions, and the laws of the land is neither possible nor commendable; it still recommends that individuals attempt to follow their conscience, but this is qualified by the admission that, at times, it is impossible to follow one's conscience. In that case modern individualist ethics recommends two strategies. First, that legislations and institutional reforms be repeatedly designed to minimize conflict. Second, that individual conscience should yield to public rules seldom, reluctantly, and in protest. These two recommendations are never easy to implement and it is never easy to know how to apply them; moreover, each by itself is not of much use.

Consider the difference between a rupture of some plumbing in the home and one in the human body. In either case, who is responsible for the detection of the damage and for its repair, the owner of the defective item or the expert who fixes them? According to the classical individualist moral principle, both fall under the owner's responsibility. Moreover, the individuals obliged to detect or rectify faults but are unable, unqualified, or indisposed to do so are thereby obliged to delegate the task to experts as legally specified. Following the classical individualist principle of delegation of authority, the expert is responsible to the individual whose authorization empowered the expert to act on that individual's behalf. By the very act of such authorization, responsibility is transferred from the individual consumer of a service to the expert who renders it. But this act is reversible and can and ought to be checked by the responsible individual. The individual initially responsible has the right to stop the expert at any time, because the expert is responsible to that individual consumer who is the expert's customer. Therefore, in societies which adopt the classical individualist moral principle in their legislation, no legal claim can be raised against an expert unless there is a customer who launches a civil complaint for a breach of contract. (Hence it is subject to the law of tort, not to criminal law. Some extreme cases are of criminal neglect and so they are obviously subject to criminal law. These cases are only technically related to the contracts involved: often, judging a case to constitute criminal neglect does not depend on whether or not a contract is involved.) Moreover, since the interaction of the parties in question is based on a contract, the parties may come to the initial negotiations with their given initial responsibilities, some or all of which are possibly delegated by the negotiated contract.

The same holds for medical systems: for any particular action, physicians get their authorization from their customers, not from society or the state. In principle, there is no difference between a physician and a plumber, a barber or any other technician. Admittedly, there may be a difference of context here, as no medical practice is permitted unless the practitioner is a licensed physician (except in extremely exceptional cases such as cases of emergency, when it is everyone's duty to help the patient). But this is true of many professions that require by law license to practice. Laws are rules reflecting and enforcing society's values. In Western democracy laws are often supposed to impose individualist morality, and thus to take care of morally legitimate individual interests. This supposition is central to a literature concerning social and political philosophy, and especially legal philosophy. This literature is forced by its very initial supposition to ignore conflicts between morally legitimate individual interests and the public interest as such. At most, this literature simply demands that no such conflict should occur. Yet it does. For example, it is a morally legitimate individual interest to avoid conscription, quite contrary to the public interest. When that literature recognizes this conflict, it declares the case a matter of expediency. The demand that private and public interest should not clash is too strong; making expediency an exception to this demand makes it too weak, since any conflict between the private and the public interest is in itself a case of expediency; and anyway life is full of expediencies.

This criticism leads to the suggestion that the public interest accords with the individual's long-range interest; that individual long-range interests and short-range ones conflict in times of expediency.

The discussion has now deviated from the problem. As long as individuals are in charge of their interests, inner conflicts, such as between long-range interests and short-range ones, are private affairs: individualist ethics never recommends imposing long-term interests on individuals for their own good.

This imposition is called paternalism, and individualist ethics opposes it in the name of liberalism, the doctrine that affirms everyone's right to hurt oneself. Nor is the conflict between private and public interest in a democracy a concern for ethics as long as the public interest is democratically determined. Yet when individual conscience conflicts with public procedure, things differ.

Individualist ethics lays the responsibility for individual action at the acting individual's door. In principle and in practice, this does not differ when applied to medical diagnosis or treatment in general or to the computer assisted diagnostic service, or to other medical computer services, such as data storage services, computerized axial tomography (CAT scan), and the pharmaceutic industry. The classical literature on individualist ethics stresses the autonomy of the individual. In order to insure its primacy as compared with the control over the individual by institutions, they declared that institutions have no interests. This is understandable but it does not help those who discuss real moral problems in concrete situations. In concrete situations all individuals--including the autonomous-- are regularly constrained by institutions, and declaring institutions less real than individuals does not remove these constraints. What does reduce these constraints at times is social and legal reform. This is possible only because the autonomous citizen may decide to be law abiding while living in a normal civil society. The autonomous citizen may and should decide to behave decently toward fellow citizens even when the law permits other conduct. For example, in democracy deceit is illegal only in specific cases, such as deceit with intent to defraud; yet deceit is almost always immoral, and should be avoided whenever possible even though it is not illegal. The citizen also may and should undertake certain social roles which incur certain responsibilities, and then decide to act responsibly within the current political and legal framework and to participate in government. This way, responsibility is always laid at the door of some individuals. In some sense, then, all social institutions are devoid of all inherent moral force: they may be granted moral force, they may be designed so as to encourage moral or immoral conduct, and they may constitute matters subject to moral concern--but not more: by themselves they can have no responsibility.

This conclusion--that social institutions are devoid of inherent moral force or responsibility even when their structures impose severe constraints on individual moral action--is particularly pertinent to any advance in medical service, mechanized or otherwise. A computer-assisted diagnostic service offering a questionable diagnosis is easy to imagine: it would offer diagnoses that a reasonable and responsible practitioner can and should reject. If that diagnosis is applied with ill-effect, the fault may then be ascribed to the computer and be left uncorrected, as no one would be in charge of correcting it. For, if there were some individual institutionally responsible for the correction, then that individual will be ascribed the error, and then some individual will be in charge of asking whether the error was culpable or not.

An institution can have an interest, in particular the interest of survival which it receives from all of the individuals engaged in it, but it is not an inherent interest as it is acquired from these individuals; it is the same case as that of responsibility: unless some of these individuals are institutionally ascribed some distinct responsibilities, the system is not responsibly designed: when an institution does possess an interest, having no individual in charge of its responsibilities, its interest will make it disclaim responsibility for its faults, and thus it will encourage individuals to ignore and conceal these faults. The usual response to this is moral indignation, which is misplaced and futile. What is needed is minor reform: once an individual is paid to be the trouble-shooter of the institution, then it will be in the interest of that individual to see to it that faults in it are not overlooked. The system, then, may be treated as a corporate individual. The problem, how can one avoid institutional irresponsibility, then, or institutional corruption, is a technical problem, a problem not inherent to any single system but rather to the (democratic) system as a whole. The mechanized and/or instituted part of any human complex is devoid of all responsibility but can, at best, be designed in order to constitute an incentive for responsible conduct. (This is why repeatedly institutions are regarded, metaphorically, as machines.) It is the irresponsible citizen who will disown responsibility by blaming the system as a whole or any part of it. For, blaming the system is evading the problem altogether, as does being scandalized by the system's inertia, by its disposition to perpetuate itself uncorrected.

Taking individualist ethics to deny moral responsibility to institutions or taking individualist ethics to deny moral responsibility even to the law (in order to render it the duty of legislators to design laws which would encourage moral conduct), we raise the question: What is to be done when ethics and the law clash? This question is commonsense, it is outside the sphere of classical (reductionist) individualism, it is central to (systemic or modified) individualism and merits detailed study.

During the war in Vietnam the popular reading of the Nuremberg trials took it for granted, and set the precedence for taking it for granted, that conduct should always be governed by individual conscience to the exclusion of immoral conduct. This is a facile solution which tends to destroy both individual and society because it puts excessive burden on every individual conscience. No field of human practice is better testimony for this kind of harmful effect than the case of medical practice. For, many cases are known in which physicians tend to act more bravely and nobly than the average citizens, and they tend to pay personally a high price for that burden: at times some physicians crush under the burden of the conflicting demands of the law and of their conscience. There is a great deal of pressure on physicians to act responsibly. There is no need to make them undertake special duties. In particular, they need never be expected to stand above the law. It follows, then, that the greater the conflict between common morality and the law, the greater the incentive for physicians to act as authorities in disregard of the law, so as to displace the conflict from the public arena to their interpersonal relations with patients.

A discussion of clashes between ethics and the law might serve as an illustration. Let us present three kinds of cases here, each of which has great significance. Between them they cover almost the whole field of the clash between ethics and the law: (1) the moral demand for legal reform; more particularly, (2) civil disobedience; and especially (3) disobedience to superior command in emergency.

(1) In civilized democracy one is obliged to obey the law even when it is immoral. In addition, one is allowed, perhaps invited, or even morally obliged, to participate in the democratic process to some extent, in an effort to reform the law so as to reduce its immorality. Ever since H.L.A. Hart published his classic pamphlet, *Law, Liberty and Morality* and the reform of the British (and other) law to be more tolerant to so-called victimless crimes (such as homosexual practices between consenting adults), the idea that the law should be morally improved has become common property; it is, indeed, a cardinal part of modern democracy though still not sufficiently fully practiced. It is generally recognized that one is obliged to respect both the law and one's own conscience. More cannot be reasonably expected.

(2) The case of civil disobedience does not violate the demand to obey the law while reforming it. Acting in civil disobedience is in the recognition of the law and of the violation of it--a recognition expressed in the very readiness to accept the penalty involved. This is usually done in an attempt to arouse public awareness of the immorality of the law one breaks and hopefully to accelerate overdue reform.

(3) Finally, disobedience to superior command under conditions of emergency. There is a growing literature on the matter, which stems from Yoram Dinstein's seminal work which suggests a simple rule: when a command is known to violate international law, it should not be obeyed. Whether a soldier under serious threat in the field or a physician witnessing torture will actually disobey a command, is a different matter; in any case, such situations pose an open difficult question which cannot be adequately treated here.

In all of these cases of differences between moral and legal demands, medicine is a special case, perhaps a difficult one, but not an exception.

To return to our discussion of morality, individualist ethics recognizes individuals as responsible within their private domain, and capable of delegating responsibility through public appointment and/or contracts, of which medical diagnosis and treatment are special cases. In addition, physicians have been traditionally bound by the Hippocratic code, which requires that they should act in the interest of their patients. This fact has led to a traditional dilemma: either the code demands what is legally/morally required anyway, or it has no right to make its demands. We resolve this dilemma by presenting some use for the Hippocratic code, as a set of demands which are not legally binding but are morally commendable. Here the conflict between paternalist and liberal ethics is resolved by systemic ethics: the paternalist ethics takes the law as the standard and the liberal ethics takes morality as the standard; we take both ethics and the law as standards, and see here both moral leeway and possible conflict between the two. As to the conflict, it may express a conflict between physician and patient. Examples for this fill the literature of forensic medicine. The standard example is of a patient who explicitly forbade the surgeon to perform anything except a specific operation, and a surgeon who has agreed to operate under these terms, and then, having anesthetized and cut the patient open, found a cancer. Under the canons of individualist ethics possibly the cancer should be left alone, but under Hippocratic code it should definitely be removed.

There is no doubt that the Hippocratic code enjoins the physician to be a Good Samaritan, to act above and beyond the call of duty. Is this not an injunction to act in a paternalist fashion? The fear that it is has caused some democratic societies to abolish the Hippocratic oath. The patient orientation, presumably implicit in the Hippocratic code is, indeed, not necessarily individualist; but as it is not necessarily paternalist, can it be maintained without threatening individualist ethics? How? We recommend that this should be done by instituting cooperation with patients through the legal requirement here advocated for informed consent in diagnosis: in that case the added effort of any physician wishing to be a Good Samaritan will have to be free of paternalism, since all possible conflict between physician and patient will have to be aired in consultation.

The concern of the present book generally lies beyond the domain of ethics, individual or social. The present book applies ethics to only one given end: increasing the efficiency of the responsible diagnostic agent. We wish to take the patient as the ultimate responsible agent and thus as the decision maker in the diagnostic process; and, within the given system, the question, whether to consult the computer, may be left to patient and to diagnostician. Any other division of functions between individuals will conflict with the principle of individualist ethics. This principle is repeatedly endorsed in the present book. Its application, however, is often problematic because of its social aspect. How should society regulate and control the proper execution of voluntarily accepted tasks? How should public appointees act responsibly and efficiently as public controllers of a partly operative and rapidly growing service?

3. THE PLACE OF INFORMED CONSENT IN DIAGNOSTICS

Thus far the discussion centered on the individualist principle of responsibility. We now turn to its application to diagnostics in general and to the study of computer-assisted diagnosis in particular. We now ask, What is the diagnostician's moral status? What is the diagnostician responsible for?

Traditionally, medical diagnosis is not regarded a place for ethical questions: allegedly these arise more frequently when treatment is considered. When physicians only advise patients on treatment plans, then supposedly the patients' opportunity to accept or reject the advice establishes the moral neutrality of diagnosis: patients are supposed to consider carefully explanations given for the treatment plan and participate in the process of decision making, and then choose among offered alternatives, or postpone or suspend decision, or ask for further consultation. Whatever decisions patients endorse, since they are the responsible decision makers, the traditional view is that this involves no specific moral considerations beyond those of any form of advice. Physicians are usually supposed to offer the best advice available, and then (except for cases of emergency) whether they agree to undertake the treatments decided upon or not, they are supposed to have discharged their diagnostic duties to the full.

This argument is valid, as we were at pains to explain in the previous pages; it applies to diagnostics and treatment alike. Yet the picture drawn in the previous paragraph is idealized; as an idealized picture it holds for diagnosis and treatment alike; when rendered more realistic, it applies to neither. Moreover, since the delegation of authority is a general matter, it is always an open question, what kind of delegation should be under legal supervision and tutelage? How much should patients know about medical diagnosis in general and their own cases in particular? To what degree should they partake in informed decision making? Admittedly, this (well-known) problem is characteristic of any system of diagnosis and of prescription, not in medicine alone: it is rooted in the principle that no one can delegate an action responsibly unless one understands some rather well-specified consequences of the action in question. All experts have the same moral obligation to explain matters to their customers. Customers may be individuals, corporations, the government or the public representatives. The explanation has to offer a detailed picture of the given problem and of the measures which are being taken in order to solve it and certain consequences of choosing them. (Cases involving technological innovation, incidentally, in medicine as elsewhere, are particularly problematic because it is so hard to specify in advance what consequences consultants should study and then describe to their customers. Also, in these cases it is often very hard to explain to the inexpert all the important details of the consequences of the choice of the innovative treatment.) Yet this explanation is required for the proper discharge of the relevant informed consent, even though it may amount to an advanced course in medicine or in any other relevant science, something for which most of us are quite unprepared--intellectually and, in the case of medicine, emotionally as well. **It is the urgency of health problems, the irretrievability of whatever is lost in ill-treatment, and the special intellectual and emotional difficulty to comprehend what one gives consent to, that are specific to medicine; and they are specific to diagnosis no less than to treatment. We therefore propose to extend the matter of informed consent to cover treatment and diagnosis alike.** This, will help heal the chief discontent with informed consent as practiced today, which reflects the difficulty of explaining to patients what they are supposed to allow.

It has been argued that the difficulty to explain medical matters to patients demonstrate that the principle of informed consent is impractical. Recently judges and physicians have withdrawn their unequivocal support for the principle of informed consent. This was amply illustrated by Jay Katz, one of the leading defenders of the right of patients for informed decision making. Katz characterizes the Hippocratic Code as authoritarian, as still prevailing and as comprising a major obstacle on the road to the improvement of medical practice with the aid of individualist ethics.

The problem of informed consent may be solved to a large extent, we say, by placing the diagnostic process under the same rules as those which govern treatment. Informed consent poses both a moral problem and a practical one. Very often practical constraints may hinder the application of moral principles. Under such conditions the constrained moral demands are too idealized and will remain so unless there is a radical reform of the context within which they are meant to operate. The change will create a more comfortable environment for the improvement of the moral quality of ordinary practice--in medicine as elsewhere. In medicine, however, this change may be planned while taking advantage of one of the most far-reaching revolutions that medicine may have witnessed: the possible introduction of a comprehensive computer-assisted medical diagnostic service.

Our claim rests on a general consideration. Traditional philosophy demanded idealization, namely simplification by isolation, both in order to get the heart of each case and in order to acquire knowledge about it. When application was bogged down because idealization was oversimplified and so unreal, the excuse was given that it is impossible to live up to ideals. This is obfuscation, both intellectually and morally: simplicity and isolation are not matters of morality. It is better to see that idealizations are false rather than to adhere to them dogmatically and be forced to a position of moral compromise. We all prefer a reduction in the level of dogmatism to any compromise, especially in matters of morals. We suggest here that often problems look insoluble because, as Karl Popper observes, no one is ever fully aware of one's dogmas, and usually dogmas close roads towards reasonable solutions. Moreover, proper choice is seldom between no-solution and good solution; usually, in the better case choice is of the lesser evil, and in the worse case, which is regrettably all too common, it is between frustration and merely opening ways to handle problems: solutions are always tentative and often fluid anyway. Classical individualism demanded too much on the basis of excessive idealization and excessive optimism.

In line with this view of choice as between uneasy alternatives, the demand for informed consent may be viewed as idealization of the choice procedure; ways to improve information flow in the diagnostic session may then be sought for. As diagnosis is performed with considerations of extra-medical factors on which diagnosticians either have prior information or elicit information from their patients, this kind of consideration can easily be the first item through which to elicit informed consent. More generally, it is best to convey information to patients at the early stages of diagnosis. What is required, then, is merely that diagnosis is performed not one-sidedly but in cooperation between diagnostician and patient. The legislation of the requirement for informed consent in diagnosis will better guarantee cooperation. This proposal is meant merely as an outline. Details have to be worked out and tested and improved upon.

Paternalism is opposed to liberalism as well as to individual responsibility and to the demand for informed consent in the case of responsible delegation of powers. It puts a burden of decision on people in authority, inviting them to decide in the best interest of others, while suspending consideration of their own interest. Why should anyone bear the difficult burden loaded on them by paternalism when liberalism requires so much less? According to Jay Katz, physicians endorse paternalism because it exempts them from openly exchanging information with patients and it thus enables them to conceal their ignorance and possible error. This may often very well be the case: the better-informed the physicians, the more qualified they are to explain themselves to patients, and thus to act in accord with individualism and thus alleviate the moral burden loaded on them by paternalism.

How should the principle of informed consent be applied to diagnosis? The question is new. Even the studies that advocate informed consent in diagnosis do not raise this question. Expert diagnosticians often behave like inquisitors, and the computer programs which are based on their performance enhance this form of conduct: cold and calculating, demanding maximal cooperation and offering none. The application of the principle of informed consent to diagnostic questioning amounts to the abolition of diagnostic inquisition, and in the following manner.

The lack of understanding needed for meaningful informed consent is due to the lack of background information, which is, quite generally, more accessible and more understandable to non-experts than specific technical details. Unfortunately, specialized education makes many an expert inept at handling it. This suggests a fairly simple procedure: patients should be offered in advance both information and explanation both concerning their right to stop every diagnostic procedure at any moment and concerning the losses and gains that this may incur. Every diagnostic session should start with a standard set of explanations. These should be ready and available to patients upon request, to justify any further diagnostic procedure, such as further questions and tests. This procedure will foster an open conversation between both parties right from the start and will prepare the stage for therapeutic informed consent. Thus, though the standing problem of informed consent cannot be fully satisfactorily solved, it might be greatly alleviated by some simple practical techniques. We will now discuss the role computers can play in the process.

Computers should provide background information to both lay and expert readers--the way many computer programs do these days. Computers might increase and enhance the initial information available; they might aid in the rational planning of patients' education as well, especially in times of great shortage of medical practitioners. Access to medical information via computers may be through different channels: pre-programmed computerized questionnaires; powerfully indexed and crossed-indexed simplified, condensed manuals of medical information; simplified tables of standard statistical data; simplified standard diagnostic trees; computerized dictionaries of medical terms; and so forth. This way public understanding would be vastly improved. The available information bank can include increasingly sophisticated material. Including frankly problematic material as a part of the publicly accessed medical information, would be of tremendous benefit, as it will raise the quality of physician-patient interaction as well as the general level of education.

The technicalities of providing background information to users of specific computer programs are covered by an already existing new specialized field of expertise, the one pertaining to data bases. The information which the computer diagnostic programs should offer patients has to be multi-levelled: it should address patients with different levels of information and circumstances; it should address patients whether they are assisted by diagnosticians or not; and it should address diagnosticians, and even diagnostic conferences. Consulting the computer service should not preclude consulting colleagues; rather it can easily raise the value of all information exchange by enabling their participants to share a broad, updated, standard information pool. Likewise, follow-up and follow-up checking processes might be represented as diagnostic problems, and so they can be partially programmed. This practice will enable a close checking of the quality of care. (The quality of care is often forgotten because usually discussions of it are conducted while focusing on mortality while evading discussion of morbidity. This evasion serves the purpose of controlling quality in the intent of containing cost rather than monitoring the quality of care.) Vital parts of treatment are thereby brought right into the center of the diagnostic process. Diagnosis does not end where treatment begins: within the medical system the two intertwine and complement each other.

This view is systemic; it facilitates the growth of informed consent as it spreads the opportunity of exchange of information and of piecemeal procurement of consent to the various stages of treatment. Indeed, the popularity of the two myths--that diagnosis is restricted to initial clinical diagnostic encounters and that diagnosis is utterly separate from treatment, is easily explicable as a mere excuse for viewing as a blanket tacit consent to be diagnosed by the very fact that usually patients initiate diagnostic encounters, and for viewing as a blanket consent for treatment the fact that they do not emphatically refuse treatment. Each of these views rests on the extravagant pretense that physicians know the whole course of any treatment ahead of time. These views are excuses, and, indeed, they`are regularly offered in response to demands for more respect for patients and for their right to decide for themselves. Excuses need not be criticized; shallow as they are, being grounds for blanket suppositions, if they work at all they work so well as to do away with all respect--for patients and for those who demand it. There is only a decision to make--between respecting and ignoring patients' (legal) rights to decide for themselves. Ignoring this right may be done with any excuse; respecting it is facilitated by the systemic approach here advocated, as it facilitates the systematic and continued procurement of patients' informed consent for both diagnosis and treatment. In a significant sense, it is trivial that diagnosis and treatment intertwine; diagnosis means differential knowledge, and the demand for knowledge does not cease after treatment begins. Yet, in another significant sense, this point is far from trivial, as its opposite is instituted in the current medical diagnostic system: as a result of the myth that all diagnosis is pre-treatment, current medical diagnosis is almost exclusively geared to pre-treatment clinical diagnostic sessions. The myth thus leads to the neglect of the exceptions to it. Here are some examples.

Our first example for the neglect of known exceptions to the myth that all diagnosis is pre-treatment is the neglect of the required periodical re-diagnosis of chronic patients which Charles W. Lidz and his co-workers have recently observed. The same, we say, holds all sorts of post-treatment checkups. Similarly, recent studies suggest that an essential and traditional part of diagnosis, namely pathology, is now suffering progressive neglect. As a result of this neglect, vast areas of diagnostic practice are regrettably ignored and not sufficiently publicly controlled. This is especially so in areas where no treatment is prescribed, for example, the whole vast and vital area of public health. Since there is no individual treatment present in this field, there is no concrete individual patient present (since concrete patients are subject to treatment as individuals, not as a public); hence, there is no individual control possible (except when it is too late, such as when poor preparation leads to uncontrollable epidemic, and through public political or administrative action).

This pertains to the ethics of epidemiology: since the advent of mass inoculation it was noticed that public health systems unavoidably play God as the decision to use or not to use any technique on the public at large benefits some individuals and hurts others--hopefully few. But as inoculations do kill, the moral problem is insoluble within the traditional system of individualist ethics and seems to invite collectivist ethics. The solution is systemic, of course, and well within the normal purvey of the democratic decision process, and the more consumer organizations are involved in the democratic decision process the better. Needless to say, the study of each problem of this sort is diagnostic though not in the least clinical.

With the advent of computer technology the problems besetting epidemiological research have become ever more acute: with a push of a button a clinician-scientist can enroll individuals for research without their prior knowledge. University hospital administrations view this as a serious legal problem and they have solved it by the introduction of forms for informed consent to be signed by every patient upon admission. This solution may be legally sufficient (we doubt it, since courts tend to disqualify all blanket waivers given in advance and under duress) but clearly it is morally evasive and socially unacceptable. Consumer groups and diverse review boards are now engaged in different parts of the civilized world attempting to regulate these matters. Now, clearly, the procurement of epidemiological data about individual patients requires no consent. What is required here is the protection of privacy. This too is handled by the same organizations and we have nothing to add to their work.

Finally, there is the current neglect of a central item in pathology: that of the *post mortem* procedures. These are clearly diagnostic procedures yet they are past all treatment. The commonsense view of the *post mortem* procedure is that it is beyond clinical diagnosis, since the patient is dead; by this commonsense view *post mortem* is either research or education, preparatory for future diagnoses. This is an error: in as much as *post mortem* or post-treatment procedures are tests of diagnoses they are diagnostic proper even when the resultant diagnoses are not preparatory to any medical treatment. That medical diagnosis may have practical import yet lead to no medical treatment is very obvious: it is so in almost all law-court-instituted medical diagnostic procedures, no matter what the purpose of the courts is, since courts hardly ever prescribe medical treatment. (A similar case is the fascinating diagnosis of persons who are too remote--in space or in time--to be prospective patients.)

The *post mortem* procedure is diagnosis whose usual function is to test previous diagnosis. This function of diagnosis is common but traditionally overlooked as it clashes with the myth that diagnosis should precede treatment. This counter-example to the myth, however, is overlooked also on the pretext that medicine ought to be scientific and science constitutes certain knowledge: the myth comes to secure certitude, and the counter-example is of unavoidable ignorance. But, ever since the Einsteinian revolution, the idea of complete and certain knowledge is no longer upheld. Even those who once preached it were also painfully aware of the uncertainties of clinical practice (e.g., Sir William Osler and Richard Cabot). And given uncertainty, the *post mortem* procedure is a part of the diagnostic process, since, especially from the scientific point of view, diagnosis is not only knowledge attained for the treatment of a single patient, nor can it be confined to any single clinical stage. As to the practical point of view, the acquisition of knowledge necessary for public control is eminently diagnostic. Computers, then, may play an integral and meaningful role in any public control system, including *post mortem*.

So much for the prevalence of exceptions to the myth that diagnosis should precede treatment and their neglect. Let us conclude this discussion by returning to principles. The individualist principle puts individual patients as the sole carrier of responsibility for their own health, and as able to delegate this responsibility responsibly to some qualified agent. Patients' comprehension of their decisions is necessary for rendering them responsible. Comprehension can be significantly increased by simple means. These means have to be instituted; since institutions are devoid of all responsibility, they, in their turn, have to be controlled by properly instituted means--with individual appointees responsible for their control. Such institutional measures are vital and should be implemented in medicine well before the systems of computer-assisted diagnosis become widespread. Though the effective tool is available, public control is no guarantee for its success; it is the sole means for maintaining human responsibility and the only means available within the democratic process.

Effective public control, like any other effective monitoring system, intensifies the interaction between diverse kinds of diagnostic sub-systems--particularly the clinical and the research, but also the diagnostic and the therapeutic. It thereby also correlates moral autonomy with intellectual or cognitive autonomy. It is reasonable, therefore, to hope that medical research will help facilitate the growth of moral autonomy in medicine.

4. THE CUSTOMER OF THE SCIENTIFIC SYSTEM

When discussing the ethics of medical treatment (the diagnostic process included) we take it for granted that the patients are its intended beneficiaries: physicians are secondary beneficiaries. This is drastically altered when patients are guinea pigs-- whether in the context of medical research or of medical training, and usually in both together. One might conclude that patients will be wise to avoid playing the role of guinea pigs. This conclusion runs contrary to empirical findings: according to *Consumers' Report* the best medical service, especially diagnostic service, is in clinics and hospitals attached to medical schools in universities--where patients are frankly used as guinea pigs for training and for pure and applied research.

In any case, individual customers of any medical service whatsoever often play the dual role of patient and guinea pig, and for a good reason: when the customer is a mere patient, learning opportunity is lost; when the customer is a mere guinea pig proper, treatment is put in jeopardy. To compromise in such a case is permissible only under stringent conditions, clearly specified by laws, whose guiding principle is that serving as a guinea pig should be voluntary and at minimal cost. It is tolerated not only because cost is minimized, but also because cost may be distributed. This is so even when payment is indivisible, such as the death sentence: in that case individuals who are under sentence of death gamble, thereby receiving a possible new lease on life; some of them succeed and the rest fail to revoke the death sentence. Such gambles are eminently rational; most research cases are considered rational on this ground, yet their rationality should be examined. To this end, it is pertinent that research--pure, basic and/or applied--should be vigorously conducted . For, the matter combines different levels of investigation, from the practical to the fundamental. There are matters of the accuracy and reliability of diagnosis and prognosis, of the legitimacy of the use of computers in the deliberations when this is the case, and the question of conflict of interests when physicians act simultaneously as practitioners and as researchers.

When clinician-scientists act as scientists, do they act as the intended beneficiaries of their actions?

Does science have products to sell? If yes, who is its customer?

Let us first look at pure science. Methodology is often the study of pure science as if it were a game. Let us first look at another game, say, chess. Does chess have products to sell? There are, of course, chess boards and pieces and other tools necessary for playing. Let us ignore these as negligible. Are there any other products to chess? The answer is arbitrary. Possibly chess provides material for news about chess, for television programs about chess tournaments, and so on. On this assumption, there is a vast market for the products of the playing of chess, or at least a market of derivative entertainment. Possibly, however, interest in chess is expressed merely in playing chess or in making the game available to players. On this assumption there is no market for the products of the playing of chess, except as instructions for players wishing to repeat a game played by others. Thus, still excluding the means for playing chess, and considering other of its products for sale, on the second assumption there are none. This is not the end of the arbitrariness. Is the sense of victory which follows a game a part of the game or not? It is, anyway, a product that can be sold, e.g., to the co-nationals of the winner in an international tournament to boost their sense of national pride. The benefits which an institution purchases from a researcher is economically similar to the national pride felt for a chess hero which the media market sells to the nation. Let us ignore all such things as pride, satisfaction, curiosity, privileges, social status, and even professional employment in a chess club or a scientific institution. Nothing then remains to be sold except for the means for playing the game which includes in one case chess boards and chess pieces and clubhouses, and in the other case university systems, computer systems and atom smashers.

Science as a tool for technology is an important commodity. Scientific knowledge reappears as a commodity, often as a means of production (in demand by producers), as part of the input into industry.

There are varied and at times complex ways in which scientific knowledge proper may be considered a commodity. A technological discovery is often viewed as an invention which can be patented. Whereas the worth of scientific knowledge is conditioned on its being publishable, the worth of a technological innovation may be conditioned on its remaining a trade secret. By definition patentable invention is a creative product that cannot be mechanically reprocessed.

As stipulated here, any activity is split into its game part and business part. The game part includes the business of supplying the tools for the game. The game is then played for pleasure, be it relaxation, enhancing self-esteem or social status, satisfying curiosity, or anything else. The business part of that activity is the production of goods and services for the market. This dichotomy is arbitrary; it arbitrarily excludes from the picture all sorts of sellable by-products of the game.

Until recently most scientists were self-employed when they were engaged in research. Their curiosity alone decided the direction of their research. But researchers were never isolated: they always used some means of communication. They had regular open debates which sometimes matured and developed into the performance of observations and experiments. Today the financing, planning, and performance of experiments are activities usually practiced well within the national socioeconomic framework. Now this is the rule, mainly because scientists are heavily dependent on sharing information and on economic support.

Why does society finance such an expensive game? First, because we play it all: some measure of curiosity is shared by all. Second, since one never knows what pure science can contribute to applied science, financing pure science is also a business gamble. Pure science is usually financed and checked by scientific institutions, scientists being its planners, customers, and producers. Applied science is financed by those who hire the scientific minds or purchase their products, be it society at large, private enterprise, the public sector, or even political or religious parties.

The aim of the scientific process in pure science is to augment understanding. The interest in pure science is usually divided into different fields and sub-fields, which are usually divided to different administrative units in universities and in research institutes. Scientists then specialize and confine their research into given sub-fields. Supposing specialization to be necessary, it still raises a well-known difficult question: Who decides which item is sufficiently important to be under specialized investigation? Is it the specializing scientist, or some scientific authority? Moreover, how and by what criteria should one decide what is important? Is it the fashionable or the exceptional? For a person who is engaged as a professional scientist and who is not over-ambitious, the easiest professional stance is that of endorsing the fashionable--and the fashionable is whatever is currently deemed interesting, important or worthwhile by the leading members of the faculty, or by peers. The fact that most scientists are driven to endorse the fashionable is now well known, and the jargon terms for them and for their scientific activity are "normal scientists", "normal science", and "puzzle solving". These terms are the invention of Thomas S. Kuhn, who speaks of abnormal scientists either as leaders, trend setters, people in authority, or as deviants who are rejects. Kuhn's views conflict with those of Karl Popper concerning both scientific method and political philosophy, namely his demand for the democratic control over experts. According to this philosophy, scientists do not simply follow their leaders; the interaction between normal scientists and scientific leaders is then two-way. The facts are nearer to Kuhn's description than to Popper's. This poses a serious question: How can we prevent the oppressive imposition of the fashion on scientific eccentrics or rebels or innovators? How can their curiosity be protected from the pressures of fashions?

As fashion may be objectionable, the right to oppose any fashion must be upheld and safeguarded. Fashions must be monitored against possible harm, by some institutions designed to check the democracy and flexibility of scientific research. Monitoring fashions cannot be computerized but computers can be used by the monitors of fashions. As long as specialization within scientific research is inescapable, it is unavoidable that specialties might be ruled by fashion; the democratic control of fashion is then indispensable. From time to time certain individuals break specialty barriers, bring about the unification of some specializations and discover new domains of inquiry that catch their colleagues' fancy. When they succeed they are considered leaders who invent new specializations and new paradigms: they may force organizations of scientific research to re-partition and to realign into different specializations.

All this is too abstract. A slightly more concrete picture looks very different. Since specialists seldom undergo professional retraining, they face the risk of getting stuck in superseded and useless types of research. They may train new recruits in the old ways and perpetuate their uselessness. This masks the re-partition of specialization. Specializations are perpetuated by many of their by-products, including library cataloging systems, computer data-base systems, and so forth. There is no easy way to curtail this since a quick overhaul of whole fields of study is not easy. Inviting people to switch from futile to intriguing studies is more straightforward, but it only works if researchers, or at least their leaders, are brave and curious; this is seldom the case.

So much for pure science. The beneficiaries here are all potential researchers, the scientific community at large, markets for the potential technological by-products and others. There is no specific customer posing demands, but merely general considerations of freedom and flexibility.

The problem is much simpler regarding technological research because the market in it is relatively free in relatively democratic countries, where the technological sub-culture is much better integrated within society at large.

The market, it is well known, is not as free as so many economists pretend or wish it to be. For example, the public control over publicly-financed technological research is both too lax and excessive. It is too lax on defunct projects financed indefinitely. It is excessive in blocking seemingly futile yet promising pioneering ones. Since the market can easily absorb reasonably risky projects, one task of the public, when financing research, might be to boost the unreasonably risky ones. This seemingly explosive issue could be reasonably dealt with by calling for criteria for selection among the very risky projects, not for their off-hand dismissal.

Finally, there is the third and middle kind of research, that sits between the pure and the technological, and is labeled by engineers as basic or fundamental research. This is pure research geared to remedy technological exigency rather than to satisfy intellectual curiosity. As technology, however, scientific research is contracted to deliver certain goods and is thus subject to the terms of its contract. Had these two sets been compatible, it would mean that basic research is run under two sets of obligations. Moreover, being pure research, basic research has no guarantees of any results whatsoever. Hence, the technological contract must be modified to take account of this risk element without penalizing researchers and without over-committing its financial resources. The agreement about stating a time-limit usually suffices for that: the financing is for a given period and in return for possible outcomes with no assurance of success.

It is no news that basic research is subject to contracts specifying limited support in return for possible outcomes, without a promise of any outcome. Nevertheless, often pressure to succeed is put on basic researchers, especially on doctoral students in the technologically-oriented bio-medical laboratories. This is so because it includes the myth of traditional inductivist philosophy according to which the success of properly conducted research is allegedly assured. Those who reject the traditional inductivist philosophy too often endorse the philosophy of normal science. This philosophy, though anti-inductivist, still promulgates the inductivist myth of assured success. It assures success as long as research is "normal", or highly conventional. Computers have been used extensively to monitor basic and technological research, but not to examine success rates and stress factors. This may profitably be undertaken.

The research to which patients contribute as guinea pigs is of all sorts. Its efficiency is often questionable, and in many ways: all aspects of social and political life intervene here and all shortcomings of modern society are reflected in its every department, including medical research. Clearly, the use of patients as guinea pigs is particularly problematic since it is a sensitive area as long as it is taboo to admit that willy-nilly patients are guinea pigs. The attitude advocated in this book is of openness and clear legal regulations which should remove both the admiration to physicians as above error and their fear of being unduly penalized for reasonable error.

Chapter 5

SYSTEMS AND MEDICINE

1. INTRODUCTION

This chapter presents the framework and method of the system of medicine, in terms of their merits and shortcomings. The course of the chapter, which consists of three sections, is prescribed by the interaction of single diagnostic encounters with the broadest diagnostic context. The first section examines the philosophical framework which could be conducive to the examination of diagnosis within a broad context. The second section presents a critical analysis of the methodological program--the systems approach--which will serve here as a heuristic device for examining the diagnostic and computer systems together. This will determine the role of epidemiology (i.e., nosological statistics). The concluding section examines the application of our methodological program to the critical analysis of the diagnostic system and the role of the computer in diagnosis.

The philosophical framework is the best place within which to present systems theory.

Traditionally, philosophy includes metaphysics, which is a set of very simple images of the universe. Traditionally, within metaphysics, two schools reigned, mechanism and holism. The holists claimed that all things in the universe are interconnected since the universe is one, and that therefore any picture of any part of the universe in isolation is an error. Hence we can know the whole universe or nothing. Holists thus divided into two traditional parties, the mystics who say one can know the universe as a whole, and the skeptics who denied that any knowledge is possible. It is no surprise that science sided with the mechanists who said every item can be dismantled to its last components, every ultimate component (atom) can be studied in perfect isolation, and systems are the ultimate components plus their interactions. According to mechanism specifying the components of a system and their modes of interaction amounts to scientific explanation.

A compromise between the two traditional metaphysical visions has been developed in the post-Einsteinian era: although the world is one system, strictly speaking, parts of it are sufficiently isolated to enable us to study them as if they were stable and isolated; a fairly accurate picture of them may thus obtain. Moreover, under different conditions the system splits into different stable sub-systems. Given conditions and relatively stable sub-systems of a given system, a seemingly mechanist theory of them may be considered adequate enough. Moreover, given different problems, the systems approach recommends tentative approximate solutions to them; using these solutions one might propose the division of the system under study (not to its smallest components, as mechanists demand but) into diverse sub-systems. In their turn, sub-systems may each be studied as a unit whose parts interact strongly; the interaction between the sub-systems may be added as a corrective to the overall picture. In different contexts, or relative to different goals, the same system may be divided differently. This metaphysical idea seems quite commonsense and its application can be easily extended with great benefit. Tentativeness, approximate solutions and conditions under which interactions between different parts are strong, rather weak or negligible--these constitute major items of the equipment of the systems approach. They turn out to be of immense heuristic force.

2. DIAGNOSIS, STATISTICS AND THE COMPUTER

Computer technology applicable to single cases is already in the market--as diagnostic expert systems software. Each diagnostic expert systems software program has to be inherently epidemiological: its programmers cannot begin to design it without reference to some nosological statistics: they have to consider first the fact that most of the prospective users of their specific product are patients sharing some given characteristics. Yet in the whole literature on the topic--which is admittedly scant and sketchy--it is difficult to trace any critical study of either the place of diagnostic expert systems in the modern medical system or of any details of relevant epidemiological factors. The value of such studies is obvious, and their absence is to be regretted. Yet the study of these program piecemeal is insufficient: as long as a minor sector of the patient population is artificially abstracted (artificially in the sense that computer programming considerations naturally outweigh other, more nosological ones), the value of any specific study will be rather limited; an integrated study of the place of expert systems in the modern market will have to be open to epidemiological considerations proper and so it will be more informative and help direct more efficiently future producers of diagnostic expert systems programs.

Likewise, public health considerations, especially epidemiological ones (when carefully taken into account), might improve the computer-assisted medical diagnostic service. This might pave the way to a comprehensive service which could tremendously improve epidemiology, if for no other reason than that it might standardize single items to enter the ensembles. This, too, has already been greatly effected by the very use of *ICD9*, which is the *International Classification of Diseases* of WHO, the World Health Organization, 9th version, and which constitutes a nosological dictionary intended for application to single cases for epidemiological purposes. The problem arises at every attempt to apply statistical results to single cases when standardization is erroneous (and error creeps everywhere) or insufficient (there is always room for further specifications to standardize). Certain test procedures and safeguards for their proper and regular implementation must be instituted. Since standard specifications never fully capture single cases, each application of statistics to a single clinical diagnostic encounter must be done through human mediation.

It is a standard complaint that statistical data are often used irresponsibly in diagnosis. The reason for the complaint is that statistics concerning a single epidemic often gain great weight in diagnosis through the disregard of other epidemics. The very implementation of the proposals advocated here, not to mention their consistent and systematic application, may suffice to reduce this undesirable phenomenon to manageable size. To generalize, many biases offer their services in the clinic, and their function is often to help physicians make decisions while in doubt and in trying circumstances. The assistance of the computer can offer more enlightenment, in helping transform as many diagnoses as possible into fully differential diagnoses, and in relieving physicians of much of their burden.

The application of statistical results to single cases is based on hypotheses concerning the individual cases, asserting that they belong to a certain ensemble. It is hard to see how these hypotheses can be tested. Yet such attributions can be tested only statistically, namely by references to samples. Hence, the only way to test them must remain epidemiological, and since epidemiological considerations are cursory and hardly ever tested, the ascription of statistics to single cases is dangerously immune to criticism: it is easy to dismiss each exception as rare; only the relative accumulation of exceptions begins to become problematic. But as statistical conclusions are taken as data, leading to more conclusions, even the accumulation of exceptions does not easily become problematic. The result is epidemiological error that is hard to detect. Epidemiologists know this, and when they suspect an error they step in and try to refute it by fresh statistical sampling. But this practice is too limited and too late. There are many examples to sustain this, and the latest examples are on public health records. The most obvious ones are false influenza epidemics which turned out to be Legionnaire Disease.

Traditionally, mechanism views statistics as probability, and probability is a measure of ignorance and/or of a degree of conviction below that of certainty. This is so-called subjectivist probability. If the application of statistics to single cases is simply the measure of the diagnostician's subjective degree of belief, then the mentality of the diagnostician influences irretrievably the diagnostic results--fortunately or disastrously but with no possible control. This is refuted by epidemiological studies of diagnoses: if it is true that young surgeons recommend intervention too often and old ones less often, then the institution of some corrective means is imperative. The subjective stance is thus refuted: it should not be regarded anymore as the empirical foundation of decisions against which there is no appeal. And if there is appeal against subjective assessments, then appeals make them increasingly objective--on the condition that the appeal itself is not decided subjectively.

The computer revolution is incomplete. It can only be completed when computers are viewed as parts of computer systems proper, handling materials in systems proper--in the case at hand, handling diagnostic systems proper and as a means to reduce subjective bias as much as possible. The way to do so must be systemic: partial epidemiology has to be anchored in epidemiology proper; the two systems--the computer system and the system of diagnosis--can together be applied to epidemiology (namely, nosological statistics) to achieve that end.

3. SYSTEMS THEORY

This section consists of four sub-sections. The first sub-section presents, however cursorily, general systems theory: as a view of the world (which is improvement on classical holism), as a research program, and as a heuristic based on that view of the world, a heuristic which allows to blend with the mechanist philosophy when studying sufficiently well-isolated sub-systems. The second sub-section presents the inter-relationships between context and system, in particular the translatability between the two, and their difference as to problems such as error-detection and precision. The third sub-section focuses on the limitations of the systems approach by applying systems analysis to it. The fourth sub-section proposes how these limitations, in part, may be both alleviated and improved upon, especially in the application of systems analysis to the systems of diagnosis and of the computer.

Systems Theory

A system is any whole which consists of parts--be they elements or sub-systems--and whose parts are fairly stable and relate to each other and to the whole both causally and functionally. The following items can be considered relatively stable and so as sub-systems: an island, a human population on that island, a village there, a family in the village, one human individual, and even one organ of that individual.

Since the scientific revolution of the 16th and the 17th centuries and until the present century, the study of systems according to this definition could not find a legitimate place in

the realm of science: science repeatedly demanded strictly
mechanical explanations relating to mechanical interactions
between elements. A mechanical explanation, we remember,
identifies the smallest elements in a system, their characteristics
and their interactions; it explains the stability of the system.
Also, in a mechanical explanation there is no room for any
goals (or final causes or teleology) and functions. Mechanism
demanded their elimination from rational science; rational
scientists were required by the mechanist philosophy to adhere
strictly to mechanical causal explanations; the expulsion of
explanations which assume functions and goals signified the
departure of modern science from the mediaeval Aristotelian
metaphysics. The success of the mechanist philosophy or of
the mechanist approach in physics has made it the main
characteristic of the rational method of scientific inquiry. All
this has led to the acceptance of the demand that all purposive
behavior should be causally explained. Since, at base, causes
needed to be linked to physical bodies, materialism--or
mechanist materialism (as distinct from the materialism
acceptable to most systems theorists)--may be presented as the
view that all systems are merely physical; materialism was
then dogmatically identified with the scientific attitude and
thus with the demand to eliminate goals. This has been
deemed the precondition for the attainment of scientific status
for any explanation of goal-seeking behavior in any field (e.g.,
biology, psychology, sociology, ethics, politics, engineering).
Against this background, it is understandable that the
introduction of the systems approach seemed irrational. This is
particularly so since the 19th century Reaction or Romantic
movement developed holism as a revolt against mechanism,
meant to be a revolt against science: systemism can easily be
identified with the holism with which it was traditionally
associated.

     To be more precise, assumptions concerning functions
and goals have never been banned from science, but have been
repeatedly reintroduced or smuggled into rational inquiry in a
tentative manner. For many reasons, their status has been left
questionable. The simplest way to smuggle functions into the
mechanist explanation is to present them in an inadequate
language of causal descriptions (the heart pumps blood, for
example) and the simplest way to smuggle purpose is to speak

of motives (such as the profit motive) i.e., of movers or of causes of motion. Yet smuggling functions and purposes forces the whole field of the study of behavior into a procrustean bed. Hence the need for a broader system for these.

One broader system was general systems theory as introduced in a very vague way by Ludwig von Bertalanffy and others. It is still a very questionable philosophy, yet it came to fulfill a strongly felt need: to break away from the mechanist procrustean bed, to speak freely of stable systems and of functions and goals. Cybernetics, systems engineering and other mathematical, pragmatically oriented practices of studies have since lent social respectability to systems theory despite its questionable philosophic ancestry. Its modern philosophical foundations are postulated in Mario Bunge's monumental *Treatise on Basic Philosophy* and his other writings.

It seems that the message and novelty of the systems approach was not the legitimation of purpose and function as scientific concepts, and not the concept of a system at large; the new suggestion implicit in the systems approach is for investigators **to study the places of sub-systems within systems.** Consider a goal-directed system. Any of its sub-systems are not necessarily subject to any specific goal. But unlike the merely mechanical systems of classical physics, it may have a function: to serve the goal of the system. Moreover, every system (except for the whole universe) is a sub-system of a higher system.

The systems approach is then a kind of program, a heuristic or a set of approaches suggesting certain methods of attacking problems related to systems integrating purposive behavior and functions as well as non-purposive phenomena (to be explained later on). These methods often lumped together under the rather mystifying title of systems analysis. As a proposal the systems approach is better viewed as a proposal to apply systems analysis to a certain kind of problems (to be explained later on). A proposal is neither true nor false, and hence it cannot be refuted, yet it can be happy or unhappy, as the current jargon has it (that is to say, a happy proposal leads to success by some given criteria), and the claim that it is happy can be subjected to critical examination and at times examining it is imperative. Naturally, some proposals have turned out to be very happy and others not.

There are many possible causes for the success or failure in any research project. Yet attempts at explanation of certain successes and failures may be undertaken, in the hope of repeating success and avoiding failure. Such explanations are of the adequacy of the successful solutions and the errors of the faulty ones. This leads to the proposal to begin with a critical examination of the systems approach. And, the severest critical examination of the systems approach should be the one conducted from the viewpoint of that very approach. The following is an attempt to apply the systems approach to itself for the purpose of critical examination.

Contexts and Systems

The mechanist philosophy requires that a system be split into its smallest parts, the study of each of its items sine context, and of their mode of interaction in general. When all of these items are thrown together there should be an explanation based on them of the system as a whole: the characteristics of the system should be derived from the idealized characteristics of its parts and their spatial configurations. (This is the spelling out of the ancient, untenable mechanist slogan, the whole is the sum of its parts. Its denial is the claim that the characteristics of a system as a whole cannot be deduced from the characterizations of its parts and their configurations alone.) The mechanist philosophy permitted the study of habitats and systems as mere stop-gaps. (A habitat is a super-system.) The result was an amazing lack of any methodical approach to systems or habitats as such, or to the contexts in which an individual part was found.

As the mechanical philosophy rejected the concepts of habitat, system and context, these concepts began to grow in the wild. The concepts of a text and a context were deemed intellectual items, or ideas; a neighborhood was understood then in the sense of some intellectual space, not in the sense of a spatio-temporal background. Consequently it was easily recognized that ideas are understood in their context and possibly misunderstood out of their context, yet it was found inordinately difficult to admit that living things could survive only in their natural habitat. The failure to realize this leads to the failure to appreciate the great breakthrough in the invention of ecology (by Lyell and Darwin). Today, with the intellectual context so altered, we find it hard to see a problem in the obvious fact that organs of living things, as even cogs of machines, operate well only within their well-ordered systems where a system is a given objective whole. When in a study of an item and its interaction with its neighborhood, reference is made to contexts rather than to systems, this at times reflects the desire to leave open the question, is the system reducible to its parts or not?

Today, when ecology is topical, it is well-known that the concept of an econiche of a given species is abstract (as a collection of those characteristics of a given environment which make it livable for members of that species), and that the concept of the ecosystem is notoriously unmanageable. The only way to handle the situation is by series of abstractions of all sorts, such as the division of the system to manageable sub-systems, and these are only possible if approximate solutions are deemed acceptable. In most cases even approximate solutions are hard to come by, but at least they have to be considered desirable before they can be found (and traditionally they were considered undesirable).

The traditional attitude to science required of it absolute certainty. Anything less was deemed subjective, partial and unsatisfactory. The demand to study items in isolation seemed to make certitude possible, yet this was an illusion: all idealized results are at best merely approximate. The Einsteinian idea of series of approximations as stepping stones has been developed by Popper, who views science as trial-and-error, as repeated and hopefully improved trials. Science, as Popper describes it, follows sets of public rules of procedure adopted by tentative decisions. In line with this view of science, a system may be divided to sub-systems in different ways, and then each division may be studied in idealization; the terms of the idealization may thereby be softened in diverse ways. This result suggests a complex network of possible approaches to any one system, each leading to a different view of it. Any option chosen, then, is a conjecture.

Any description of the context within which a system appears is likewise conjectural. One starts with a problem to which one (tentatively) ascribes a context. The context is then divided (tentatively) into the system, within which the problematic items strongly interact, and its environment. The move from the system to its context is the outcome of a shift of problems. The new problem then invites a new systems analysis. Then the same process of analysis of the context develops as the one spelled out in the previous paragraph: the context of the initial system is now the new system. Hence, the context has its own context. The question that much of the philosophical literature is concerned with these days is, can this process end with a contextless system? Can there ever be a text without a context? The tool with which one moves from a system to its context is systems analysis. This process of systems analysis, however, never spells out clearly all the decisions that go into the determination of the boundary of the context. Thus the move from a system to its context is problem-ridden, and the reiteration of this move is increasingly problematic. Hence, even if a contextless system exists, even if there is a text without a context, we are far from ever approaching it. (The contextless text would be the complete theory of the universe, of course.)

No system is ever analyzed to its sub-systems fully and in all clarity. Consequently some areas of each and every problem-situation are less clearly examined (in the jargon they are called softer) or more clearly examined (the harder parts are those more clearly analyzed). In addition, different areas of the context are more clearly examined at different attempts, depending on the varying focus of the analysis. Since not all decisions on these kinds of division are explicitly stated, the systems analysis may run the risk of unnoticed contradictions, especially in its softer areas. This is particularly so when two conflicting decisions in the above-mentioned complex are confused. In a formal system, such as a computer, a contradiction which is used in a deduction causes a breakdown in the whole system in which it occurs, since a contradictory text includes every possible statement as its logical conclusion. How are contradictions to be handled?

Formally speaking, all contradictions are equally deadly. Yet in practice, their damage is limited. Even in the most formal context of the computer their damage is limited, since some contradictions in the formal system it employs may be beyond its reach--in practice and in principle (Gödel's theorem). In normal circumstances some contradictions are easily recognized as mere misprints or slips of the pen; other contradictions are judged inherent to the system. The proof of this is the fact that referees are deemed hostile if they consider marginal contradictions in a research report to be inherent and dismiss it on that account.

How come distinctions between inherent and marginal contradictions are so often made with ease? This question should not be taken to mean that it is always so. The seeming contradiction proves this point, since it may take great efforts to decide that it is no contradiction; similarly, the question of the cost of the removal of a contradiction from a system not always easy to judge as the history of mathematics shows; similarly, this history shows examples of the discovery that a contradiction deemed marginal reappears regularly, suggesting that it requires serious attention.

How then is it judged so often with relative ease, whether a contradiction (or other error) is marginal or central? The answer is that familiarity with the context at times helps judge such things reasonably well, and at other times not. This answer is not satisfactory, but it does report an empirical fact. Can this answer be formalized and thus computerized? The answer is, as ever, that formalization can be done in part: the system is open-ended but parts of it can always be formalized (never perfectly but) to varying degrees of adequacy. In order to formalize the text a sufficiently formal meta-text is needed: systems which are possible candidates should be placed in the para-text and permitted the attempt to render any item in the para-text the tentative context of that text.

This point is of a great import for the somewhat more formal study of diagnosis (as will be illustrated later on) as the process of pattern-recognition. Quite intuitively considered, any attempt to place symptoms in different clusters of diseases is permissible to begin with, for example, where clusters are presented alternatively as nosological, ecological, or even sociological possible contexts. The varied candidates are the diverse possible systems; therefore, no reading of the text in the given context is perfect. Moreover, the system is fraught with severe lacunae, with errors and contradictions. Any attempt to examine any text should therefore be allowed to end with a failure.

Often the evaluations of a contradiction can be guided by assessments of common errors due to prevalent old-fashioned views, to contradictions, to the endorsement of new views without rephrasing ideas in use, so that some ideas reflect the new system of thought and some the old. This repeatedly happens in the case of (physicians') ignorance of the theory which serves as the context and/or meta-context.

All this illustrates the usefulness of systems theory for the theory of contexts, the centrality of the idea of approximations and of tentativeness in this nexus. The key idea is that of a sense of proportion, of the willingness to benefit from a sufficiently good approximation to the truth.

Systems Analysis of the Systems Approach

The systems approach is known for its applicability. It is the search within wholes of the inter-relations of sufficiently isolated and stable sub-systems and of elements subordinated to the purpose of the whole and functioning within it. A system for systems analysis, namely, a set of rules of thumb, has been developed and suggested, facilitating the finding of goals of given problematic systems under study, and functions of sub-systems, as well as of inter-relations of sub-systems and of parts. As any systems analysis presupposes goals, it is often technologically oriented.

Consider the task of a systems analyst in those simple cases for which systems analysis is reputed, namely, of improving the functioning of a technological concern. The first task of the systems analyst is to find the task of a given system, such as a technological organization. This is usually done in accord with demands of a customer who presumably has invited the systems analyst and whose interests are served by the system. This is a limitation on systems analysis. Other limitations exist and constitute the boundaries between systems and their environment. The environment of a system is not necessarily physical: it may include all sorts of restrictive conditions within which the system functions. Decisions concerning boundaries between systems and their environments may include bias and arbitrariness. Next, the systems analyst may have to discover who are the decision makers in the system and what kinds of changes in the measures of performance of the system they can produce. Since the function of the decision maker is tightly related to the goal of the system, the systems analyst may have to identify the decision maker early in the investigation since the whole process of systems analysis may be constrained by the decision makers of the system under investigation, by their own views and values of the and by the constraints put on their power to decide. The system may include a hierarchical system of decision makers that calls for examination. The cooperation of this sub-system ought to be enlisted for the process of systems analysis, or else implementation of proposals for change become utterly futile: no significant alteration within any system can be effected single-handedly. Thus, systems analysts profit from interaction with decision makers on diverse levels. Decisions makers will also benefit from a general picture of the system, and so it is in their interest to cooperate with systems analysts.

Viewed from this perspective, the pivotal question is whether the customer and the system's decision maker cooperate and what is the system's minimum level of performance that is acceptable to the customer. Once this minimum is identified, the value of the system to the customer can be assessed and rendered operational. These measures, in their turn, influence the function as well as the structure of the system.

Given the system and its environment, the systems analyst may examine what the system can do maximally, its performance measure, so called, as an objective fact often not given to change by the decisions to be made by customers. At times, however, as the system's decision maker or through the cooperation of that decision maker, the systems analyst may recommend to change it, improve its performance measure, or close it down, and so on, to the extent that the goal requires and the environment permits.

All that seems quite straightforward (for more detail the reader is referred to the *Selected Bibliography* at the end of this volume). Yet the question here is whether the rules advocated by systems theorists are as generally and as fruitfully applicable as they claim.

Bertalanffy has suggested in his popular *General Systems Theory* that his theory may be derived empirically and intuitively by tracing various patterns of systems throughout the multitude of natural and abstract systems. This is traditional inductivism or extreme empiricism and it clearly runs contrary to the systems approach. Were the inductive method of searching into patterns possible, were it able to bring us nearer to the general structures of systems, then there would be no need for systems analysis ever. The generalities offered to the systems analyst are to be taken as heuristic--as schemas for generating and testing hypotheses with an eye on their application. This idea competes with the idea of induction, as these schemas are not authoritative in any way: they should be subjected to critical examination.

Systems analysis as such is a system of inquiring into or of analyzing any given system with the aid of the kinds of questions which systems analysts usually ask. What is the aim of the system (of systems analysis)? How efficient is it? What are the functions of its sub-systems? Who is the customer? Who is the decision maker? What are the minimum standards that the customer demands of the system and can the decision maker achieve them in the existing environment? (Bertalanffy's suggestion to study all available systems, for example, is the abandonment of the sense of minimum standard and is thus not systemic at all.)

The goal of systems analysis is quite regularly described in the literature and may be considered as standard. According to the standard description of the goal of systems analysis, it is dual: to ascribe goals to systems, and to analyze their performance in the light of which their presumed goals be analyzed.

This description is satisfactorily applicable to systems analysis only when the aim of systems analysis is considered to be the mere acquisition of knowledge for purely scientific purposes. The intent to improve on given systems--scientific or technological--is the adoption of some recommendation to implement some proposals in the light of some findings. Hence the above description of the aim of systems analysis--though standard--is lacking in that it does not address the problem of change, including the possibility of changing goals, functions and structures of sub-systems, as well as of the system at large (while implementing some proposals based on the original study). Change usually occurs in different parts of the system, ranging from high-level more abstract systemic structures to low-level particular items of detail. The traditional approach to systems analysis has not paid enough attention to such issues, though recent developments in the field aim to rectify the problem by introducing means for objectifying and monitoring the process of change. The application of such innovations in the field of computer systems for medical use is non-existent.

The whole discussion of the systems analysis of systems has been lacking thus far in technology. Systems analysts tend to become designers of improvements, to be endorsed or rejected by the customer of the service of the systems analysis. The aim of systems analysis, then, is to help improve given technological systems by publicly prescribed standards.

An interesting theoretical problem ensues: do systems objectively possess aims and functions, or are they endowed with these by the individuals who steer them?

This question is new. Traditionally students of social affairs either denied that systems exist--they were individualists, atomists, mechanists--or else they assumed that systems, as well as sub-systems, are characterized by their very functions--the collectivists, organicists, holists and Romantics of all sorts. Nowadays, whatever is assumed about reality, and whether or not it is claimed that in the final analysis all organizations are plainly machines, it is intellectually as well as pragmatically useful to assume that systems and sub-systems exist, yet not to ascribe to their structures and functions any characteristic that binds them uniquely together. Even in the most basic biological studies of function, in physiology, it is nowadays taken as obvious that an organ, as characterized by its structure, may be given different functions under different conditions, and that different structures may serve the same functions under different conditions. It is also taken for granted that the organ's change of function may bring in its wake a change of the organ's structure and *vice versa*. There are important analogues to this mutual influence of structure and function in systems analysis.

Do systems, then, have aims in addition to their ascribed functions? Evidently some do some do not, as individuals and their organs testify. Do institutions have aims? Certainly some institutions, such as the institutions for higher learning, do have aims, endowed to them by society. Does society have aims and if so which? The individualists say no, the collectivists say yes. We side here with the individualists without debate (since the matter is examined elsewhere; see *Selected Bibliography*).

Systems analysts are seldom, as yet, invited to help plan a normal, medium-size technological system. Though systems analysis is only a few decades old, it has already a tradition, a tradition of the anticipation of a crisis, if not the meeting of a crisis in full acuteness. The requirement of sophisticated large technological systems are different, more to do with planning than crisis-control.

What is the cause of a crisis? This cannot be decided in the abstract. Yet, abstractly, one possible cause of a system's crisis may be conflicting aims of a system or conflicting aims and functions between sub-systems or even the loose coordination between the functioning of diverse sub-systems. (This last defect, incidentally, was initially considered the distinct subject-matter of operational research, but this distinctness was soon lost as systems analysis incorporated operational research early on.) How is choice made between conflicting aims? Of competing sub-systems, which should become dominant? In the psychology of character, this question is posed for every person paralyzed by inner conflict, and for those burdened with conflicting functions--i.e., habits or patterns of behavior. Yet in the case of an individual person, the right of that individual to exercise choice to the utmost governs the answer to this question. In the case of organizations, the parallel with individuals is complete only in the organization in which decision is reached by one individual, the system's single decision maker who fully or almost fully commands choice. When the system is a democracy, or even when there is democracy or power struggle among its leadership, the problem differs. It differs to the utmost when it is run by autonomous or semi-autonomous teams and when its general frame is under democratic control.

When systems analysis evolved, it pertained either to governmental and military structures or to all sorts of autocratic organizations. Some systems analysts did hit upon the idea that a major conflict-resolution technique is the opening of free communication between various teams within a given system, hopefully including the voicing of complaints and the encouragement of the proposal of resolutions. The more democratically disposed the systems analysts were, the more they encouraged the study of the system or at least the sub-system, by its own membership. This process of the system's self-study contributed to participatory democracy as well as to the democratization of the work place and the reform of work by the creation of semi-autonomous work teams. These changes are still very young and, though they are of general interest, this is not the place to discuss them. The democratic ethics they reflect is sufficiently apparent.

Leaving aside inner conflicts within a system or between its diverse sub-systems, there remains one more conflict: that between the system and the systems analyst requested to diagnose and offer cures for its ills. How serious is this conflict? At times it is unproblematic. Systems analysts aim at pleasing their customers. So, even when they design and implement some moves, by default, they aim to please. But this can happen only in simple cases. If a system is complex enough to have customers (say shareholders of a company), decision makers (management) and sub-systems (teams of workers, especially semi-autonomous teams), the situation is drastically different. And the diagnostic system in any medical complex of necessity is quite complex. In this category systems analysts are bound to be caught in a conflict of interests that they are unable to analyze, as long as they are unaware of their having overstepped their traditional functions, and when on top of their analyses they try to influence the choice of proposals and, at times, even implement them.

Authors who write about systems analysis must be aware of conflicts, yet they minimize their influence: they assume, at times explicitly, that "the designer's intentions are always good" (C.W. Churchman). This assumption may justify for these authors an idealized presentation of systems analysis, one denuded of its context (decontextualized). Yet the price is often irrealism. (It is thus no surprise that past efforts to apply systems analysis to medical complexes were not analysis successful.) Systems analysts need as much room as they can get for maneuvering in cases of conflict. Their problem is quite general, and is as follows. Part and parcel of all the problems hitherto discussed is the problem of how to ascribe sub-systems to a given system. There is no steadfast rule for the process. The systems analyst might legitimately and at times profitably choose to view any part of the system or any sub-system as dominant, and the other sub-systems as subservient to it--provided that the choice is tentative. This decision gives way to many interesting possible combinations and it influences systems analyst's decisions concerning other specific constituents of the systems they study--say, their aims, their decision makers, the right answers to the questions, who their customers are and what is their respective environments? There are no hard-and-fast rules about how the systems analyst (or the designer) should act, except for the rather general pragmatic guideline.

Here is the standard problem of the systems analyst in the midst of a conflict. Success is judged by aims. Which of the conflicting aims is the one by which to judge? This is an abstract question. In concrete cases, especially in crises, survival is the supreme end. But which of the sub-systems should survive? The general views and values of the society which houses the system are usually taken as guides. This practice secures that conflicts of interests in all practical cases of systems analysis are resolved in principle almost at once, thus enabling analysis to proceed. (Here the conflict is resolved in a soft manner, often with no awareness, and hard thinking goes to subsequent considerations. This fact is a severe limitation on systems analysis. For the purposes of this book this limitation is too expensive: it must be removed. This is the rationale of the present chapter.)

What happens if views and values do not offer sufficiently clear-cut guidelines? Perhaps the whole society is in a crisis. Systems analysis can then operate, if at all, only on the assumption that the survival of the whole society is the supreme value to uphold. Otherwise the systems analyst has given up all hope or all interest. If survival is taken as a supreme value, yet a clear-cut guideline is still not found, there is a need for a lot of commonsense and for real statesmanship. This book is not the place to further analyze this very interesting case. It might be noticed, however, that the less a problem is context-dependent, the harder it is to decide how to approach it.

The last paragraphs raise the annoying general question: if the field of systems analysis is so limited, what is the use of it all? Is the systems approach merely a red herring? No. The set of general concepts of sub-systems, customer, decision maker, environment, and so on--even though relatively highly open-ended--have been quite usefully applied to some systems, especially simple technological ones. Even if the only benefit from it is the awareness of open-endedness and thus of tentativeness, and the awareness of the value of approximate solutions, then the profit is immense. The development of cybernetics in general, and of computer systems and computer techniques in particular, have instigated fairly objective rules (such as the rules of cost effectiveness), for assessing options explicitly and weighing them rationally. The choice between the options is otherwise not open to rational treatment. This success is highly important, however tentative and limited it might be. The limit of usefulness of the systems approach is met when one tries to fix a norm for the applicability of systems concepts to more complex affairs, particularly to systems without elaborate contexts.

There is no way to decide, once and for all, as to what the aim of a given system is, because aims change in time and in accord with various possible points of view. Aims and decisions have always been a source of controversy which regrettably people tried all too often to evade by turning to authority and/or by theoretically explaining away all purposes. Systems analysts are often taken to be authorities which replace decisions by mathematics, especially when hired by faltering managements (at times also to hospital managements, regrettably) to approve of decisions already taken and lend respectability to questionable practices. Nevertheless, as long as it is remembered that no aim is intrinsic to any system, and that the boundaries of a system are variable and partly arbitrarily-determined conventions, it is possible to remain undogmatic and leave assumptions and proposals open to criticism and to monitoring before, during, and after implementation. There is no substitute for public control and for the normal democratic process. The attempt to improve this by the institution of a technocracy only imposes on the technocrat a conflict of interests. It is both in the public interest and in the interest of the expert technicians that they be subject to democratic control within their own organizations and by the public at large. To this end they have to be subject to criticism, and to that end they should not be penalized when the criticism is valid: penalty is adequate at most only in cases of gross neglect.

Regrettably, this point has been repeatedly overlooked by writers on systems analysis who are all too often (undemocratically) technocratically-oriented. Monitoring should be part of the system's inner structure as well as of its environmental function, in order to prevent stagnation, petrification and extinction. By efficient monitoring design, customer and decision maker might be in regular communication and interaction in an easily accessible network and check each other. The efficiency of the monitor itself is a simple matter given to systems analysis. Principles of democratic rational planning might be usefully applied to the systems approach.

Cost Effectiveness

The heart of systems analysis is reached when cost is discussed--financial or any other. Examination easily shows that different proposals for plans and/or alterations incur different levels and sorts of expenditures which cannot always be compared since they are of different proposals for implementations: the cost of minor changes is generally smaller than that of thorough treatment, yet at times the one is preferable, at times the other. The general rule for preference must relate cost to benefit. It is not cost that matters but cost benefit.

When a consumers' organization recommends one of the many alternative implements in the market as best-buy, it is neither quality alone nor price alone that determines the best-buy, but, as they say, whether purchasers get their money's worth. Yet, as the elementary course in economics teaches, it is not money's worth that is at issue but the worth of the purchased item for the shopper. Consumers' organizations seem to eliminate the consumers, but only because they replace them with a dummy constant--the average shopper. This procedure is explicable by a systems analysis of the consumers' organization. It may be seen that the membership of the consumers' organization play a role to its customers; the dominant sub-system is the one whose function is to provide the material that enters the news sheets of the organization; this sub-system includes, more or less, paid functionaries and officers of the organization; they aim at serving as many customers as possible; to that end they address as large a public as they can reach. The cheapest means of computing ways to achieve this is to assume the existence of an average. (To do this the sub-system includes a statistician.) This is an example of a standard systems analysis of a standard democratic, but not participatory, goal-oriented system. (Consumers' organizations promote participatory democracy; yet they are not very participatory. They become increasingly democratic as their members partake increasingly in their major decisions. They become increasingly participatory as they improve the forums which they provide for their members to exchange opinions.)

The system of medicine is similar though more intricate. As long as customers behave in a non-participatory yet rather democratic manner, it matters little whether the treatment is offered by a person distinct from the diagnostician. For example, a more actively participatory medical system would require of surgeons to check for possible conflict of interests in all proposals for major operations which they make. The more active the public participation, the more obvious it becomes that decision making and implementation are better put in different hands. Likewise, monitoring standards of performance for the use of medical--pharmacological, surgical, and computer-assisted--technology should be handled by public bodies.

Now, the comprehensive computer-assisted medical diagnostic service system open to public inspection here recommended, enables anyone to check for information and interests, and to increase thereby the comprehension of medical problems. Consequently, the recommended solutions within their complex contexts can also be better comprehended. This example is of a standard systems analysis of a standard actively participating membership of a democratic goal-oriented system; this is why systems analysis centers on the claim that patients ought to be viewed as autonomous decision makers. This makes for a diffuse system that cannot easily be broken down into sub-systems, one whose members are represented adequately for purposes of implementation with the aid of a fictitious average consumer. This fiction is essential in

epidemiology which is geared to serve the public interest; it is
not acceptable within diagnostics geared to serve individuals.

The choice of individual goals is both cumbersome and
highly problematic. Adversary relations between individual
customers and suppliers of advice thus must take place to some
extent; to that extent adversary relations are inherent to this
kind of system, and the adversity increases with the variance
from the average. Even the supply of tools for diagnosis,
concrete (machines) or abstract (procedures; epidemiology), are
geared to the average. Claiming that consultants normally
have their customers' interests at heart is thus only an
approximation. When the consultation is medical, the non-
average customers who do not get the best advice possible will
not be comforted by the knowledge that their cases are
exceptional. In adversary relations some compromise may--but
not have to--be sought and found. The obvious compromise is
to split the set of customers into sub-sets and average each
separately with the aim of reducing variance. This, as any
compromise in matters of life-and-death, is cruel. Is it
defensible? Only as an expediency.

The ideas here adumbrated may raise association with
events that currently take place. The cruelty of the situation
may, at first, invite efforts to overlook the harsh details; yet
this is sheer negligence. Systems analysis proper requires that
such analysis be spelled out in detail and implemented with
foresight and care. One detail which is regularly misperceived
due to the absence of a patient analysis of this cruel situation
is as follows. It seems that the rich can afford to ignore all
that is said here about consumer organizations, including
medical ones, because their fortune and the liberal structure of
medicine in Western society permits them to deviate from the
norm and to try to purchase superior services in the open
market. Not so. The error of this rests in the oversight of
the notorious need of the rich for control, especially in such
markets that are not easily controlled and vociferously refuse
control. Markets in which large flows of funds take place
regularly invite irregularity; producers of goods and services in
quality markets also tend to cater to the average, and only the
tailor-made good or service escapes this trap and even then
only approximately so. Today even the rich do not have every
good and service tailor-made, and contrary to appearances even

quality medical services are often rendered with too little concern for individual care. Often services are tailored to cost rather than cure.

The fear that the introduction of the computer to medical diagnosis will reduce quality is the fear of still further standardization. This fear is justified, but not the proposal to meet it by banning computers. This move incurs a tremendous cost and a small benefit: standardization occurs due to market pressures alone. A more cost effective way is to use computers in order to raise public readiness to participate, especially in the diagnostic process. The relief of the diagnosticians from some responsibility may reduce market pressures and enable participants in the process to raise the level of personal concern in normal diagnostic service.

This kind of concern should guide the development of the comprehensive computer-assisted medical diagnostic service, but first attention should be paid to the theoretical considerations which introduce cost-benefit considerations as they apply to diagnostics. It is proverbial that life is infinitely more valuable than money, and that hence cost benefit in principle does not apply in matters of saving life. It is very important then to show that this is an error: right or wrong, we do apply cost effectiveness considerations in matters of life and death, and making them more carefully and knowingly may save lives.

Chapter 6

DIAGNOSTIC THEORY

1. INTRODUCTION

In this chapter we focus on the systems of medical diagnosis. The systems approach stresses the context of the diagnostic encounter, or rather the different arenas in which the encounter occurs: the epidemiological, the socioeconomic, the social, the ethical, and the legal.

The general framework and foundation for our discussion should by now be familiar to the reader. The diagnostic encounter is partly subject to rules and partly open-ended. This combined approach allows for the use of rigid and formal rules, and is therefore capable of computer simulation, but maintains open-endedness. Sooner or later the simulation breaks down or, to use the jargon, it "cracks". This situation arises, in part, because the context may not be sufficiently fully specified and thus the simulation "cracks" when it transcends the intended boundaries of the rules. There is no way to illustrate this, except by a specific case of simulation and its having succeeded for a while until it "cracks". When a simulation "cracks" it can be mended *ad hoc*, and often with ease. What cannot be avoided is the threat of a "crack". The computer simulation may thus be improved by better specifications of contexts. It is also possible that the simulation breaks down or "cracks" because of the open-endedness of the rules implicit in the recognition of their partiality and insufficiency. In this case, the rules are ripe for revision.

The advantage of this dual approach, of using the formal and the ordinary mode of speech side-by-side, is threefold. First, the usefulness of the rules within their domains of applicability is increased, and this increases the usefulness of all computer services as such. The part of diagnosis subject to rules is the one where the computer becomes handy since it may be programmed for a successful simulation. Second, computers can help expose the weakness of the rules. For example, a computer simulation may appear adequate and the computer nevertheless recommends a clearly unacceptable diagnosis. Reform is then possibly called for, depending on cost-effectiveness considerations; computers may help expose the weakness of the rules and they may be used as a means to reduce the costs of the reform thereby invited. As examples one can choose any program that has "cracked" and was fixed with some modicum of success. Third, the recognition of the significance of the difference between the formal and ordinary modes of speech is very useful: a great impetus was given to medicine as a whole by the recognition of the need to monitor every formal system for its limitations, especially such a powerful system as the computer, particularly when applied within the domain of medical diagnosis. The custom of instituting monitors is spreading and it raises efficiency quite significantly.

The cost effectiveness of a proposed reform is a significant factor and points to the difference between science and technology. Science can be idealized as a game with players who can do no wrong; their search for the truth is unconditional and unqualified (except by the laws that govern the conduct of all citizens, of course). When the players cease to abide by the rules of science, either they leave the game (as do chess players who cheat, since chess-with-cheating is a game different from chess), or they argue that the rules constrain their search (as the inductive rules do). Technology differs radically. A criticism of a practice does not usually lead to its abolition except in the rare cases (e.g., blood-letting) where the very abolition of the practice constitutes progress: usually it invites a reform of the criticized practice.

The tremendous advantage of making all this explicit may be illustrated by exploring the significance of the null case. The expression "null case" is a derivative of the expression "null class" or "null set". Take the class or set which has no members, such as the class or set of all unicorns, or the class or set of all square-circles. Is it a class or a set proper? This expression, "null class" or "null set" is itself a derivative of the expression "null" or "zero". Is zero a number proper? The answer is, no: zero is not a number like any number. Proof: a cake can be divided between five, four, three, or two people; it is questionable whether it can be divided between one person and it certainly cannot be divided between zero persons. But zero is introduced into arithmetic so as to simplify its formal representation at a cost: often divisions by zero have to be excluded explicitly.

The null set is also admitted for formal purposes: we want to avoid having to insure that a set is non-empty before speaking of it. Similarly, the null case is the case which is or turns out to be no case at all: it is not a fictitious case, it is even a recurrent one, appearing in clinical diagnostic encounters with healthy individuals as customers of a clinic, exposed to diagnosis or not. It plays a role in diagnostics akin to zero in arithmetics and the null class in logic and mathematics: it helps present things formally, and this helps state things explicitly and clearly. The explicit clear presentation forces one to introduce from the start factors usually ignored in diagnostic theory, such as the importance of the epidemiological factor in the context of the clinical diagnostic encounter and the importance of the null case in epidemiology.

2. DIAGNOSTIC THEORY

Diagnostics must contain medical theory as a sub-system. For one thing, systems analysis requires the study of a system's performance measure, whereas theory has none. A performance measure is a criterion to measure the product of a sub-system, and it is used to help its decision maker decide what is the minimal level of efficiency the sub-system must maintain. For example, the performance measure of a hospital will depend on a few factors, including the optimal number of beds it should occupy on the average. One may then compute the number of beds below which the hospital should be closed. The financial aspect of the matter is rather easy: if too few beds are occupied so that it is cheaper to house patients in a hotel, then the hospital should be closed. Other factors, however, concern the case with which instruments may be installed in the hospital, as well as other issues, and these may radically alter the picture. Yet the question remains, How many

patients should be served on the average, with what rate of
relief and of cure? All this is a matter of technology. As a
matter of research, medicine looks quite differently.

Viewed as pure science, medical theory is not given to
the standard treatment of systems analysis and as pure science
medicine has no patients. The problems pure medical science
does have are hard to solve. There is no way to assess the
rate of scientific progress. When science is applied, however,
the system in which it is applied may have a proper
performance measure. Indeed, this helps decide under which
conditions it is advisable to apply which theory. When no
application of theory is specified, one can only speak of the
performance of researchers. Can the efficiency of research be
studied? Not at present.

The theory which researchers apply is called
methodology: the theory of scientific method. Within
methodology, science is presented as sets (or collections) of
hypotheses and sets (or collections) of empirical data, and their
logical interrelation. In particular, sets of data relate logically
to sets of hypotheses. Each set, of hypotheses and of data,
may be broken down to sets of competing sets: competing
hypotheses compete as deductive explanations, and competing
data compete as reports of the same facts; each competitor is
then assessed as to its relative degree of success, which depends
on degrees of precision, of explanatory power and other factors.
Finally, only some of the data are reported--as observed
repeatedly. The logical relations between the sets, of different
hypotheses as well as of hypotheses and data, are deducibility,
independence, and contradiction: the explained hypotheses are
deducible from the hypotheses which explain them; competitors
contradict each other; independence is neither deducibility nor
contradiction. Similarly, given data may be explained by given
hypotheses, and then the data are deducible from the
hypotheses; otherwise they are independent of them, or
contradict them. All this--the deductive part of science--is fully
agreed upon these days by all students of methodology. What
they disagree about is the matter of induction or confirmation,
which here is consigned to the sub-system of applications.

The aim of theoretical science is the following: develop
a set of hypotheses which produces as much data as possible,
which contradicts no reported data and which conforms to a

given intellectual framework (which is usually metaphysical). When some hypotheses do contradict reported data, yet the data that can be deduced from them sufficiently approximate the reported data, then the stage is set for the preparation, if at all possible, for observing more precise data than were previously reported, and then a crucial experiment or observation is performed.

So much for methodology in general. Theoretical medicine, in particular, offers a very simple intellectual framework. The framework should house hypotheses concerning diseases, syndromes, symptoms and functions. The theories describe structures and their functions, or they describe adjustments of functions to defective structures and *vice versa*. They may also determine the limits of deviation of structures and/or functions from the norm before they count as disabilities and/or diseases; these deviations are at times specified as etiologies. A set of symptoms is a syndrome, syndromes relate to a disease entity and disease entities relate to etiology. A symptom may transform into a syndrome and a syndrome into a disease entity by the use of the epithet "idiopathic". An idiopathic symptom is one with no known syndrome, disease entity, or etiology to relate to. For example, fever of unknown origin, F.U.O., has no accompanying symptom and is thus idiopathic. Finally, the theory contains sets of hypotheses concerning recommended treatments for symptoms, syndromes, diseases and etiologies, which hypotheses assert degrees of correlation between a specific treatment and prognoses, and estimates concerning the likelihood of degrees of return to average health.

Informally, this means something fairly intuitive, as an easy example will show. A sore on a penis is a symptom. In terms of a syndrome it may be diagnosed as primary syphilis. The disease entity is syphilis: primary, secondary or tertiary. The etiology has to do with the presence in the body of a micro-organism called *Treponema pallidum*.

So much for methodology in general and regarding the intellectual framework of medical theory in particular. **There is no possible performance measure for science, only for technology, since the future development of explanatory theories is in principle unpredictable.** As to the assessment of theoretical success, there are no strict rules for measuring such success. (Even if success of a theory is measured in its successful passing of empirical tests, the difficulty remains: When has a theory passed sufficiently many and severe tests to count as successful?) As to failure, there is no criterion as to when performance is so poor to be utterly useless.

The situation is different in the practical systems or contexts in which science is applied. Briefly, the application of any theory may be a private affair, guided by ambition and conscience alone, and it may be a public affair, guided by rules and regulations, by law and custom. Historians of medicine tell us of physicians who experimented on their own bodies with new modes of treatment, considering this practice noble as a matter of course, as if it invites no discussion. The stories then suggest that nobility has vanished--presumably because of the general deterioration of mores. Not so: the increased awareness of the irresponsibility and illegality of this practice has drastically discouraged it. We shall not dwell here on therapeutic technology; it looks as if there is no parallel problem in diagnostic technology, yet this is a serious error.

The process of introducing theory to the diagnostic situation involves both physician and patient. In every diagnostic session initial decision should rest with the patient-- the decision to listen to the physician. Final decision should also rest with the patient, who has to endorse or reject the physician's practical proposals. This is what individualist ethics requires, not what happens in practice; decisions are today still largely in the hands of diagnosing physicians. Sifting symptoms, gathering signs, sorting out part of the theory deemed relevant, considering contexts in many ways, all enter the physician's deliberation during the decision-making process. Moreover, the physician tries to conjecture and test a diagnosis, perhaps also an etiology. Finally, it is customary that the physician recommends treatment and explains it to the patient to some degree in order to secure a sparse informed consent. After that, prevailing wisdom dictates the (erroneous) view that diagnosis is over and that treatment begins as distinct from diagnosis. It looks as if the right to introduce theoretical innovation does not enter the process.

The above description of the diagnostic process and of the way medical theory enters it is traditional, yet it is too idealized for the present study. As we have already mentioned, it is here that methodologists disagree among themselves: they disagree more about the question of applied science than about research. Fortunately, of our present practical purposes the disagreement in question is easy to ignore as it is too idealized to bear any relevance to practice. In practice, especially in medical practice, law and/or custom require a corroboration of the claim that a practical procedure is beneficial and its side-effects are not too objectionable, as a condition for permission to apply it or to put it on sale in the open market; this corroboration has to conform to certain rules which are also legally specified and/or socially accepted; even the test procedures often require prior approval and public supervision; and the authorities which are empowered with granting the required permit for these are specified by law and/or custom.

3. RESPONSIBLE DIAGNOSTIC TECHNOLOGY

To conceive of diagnosis as a mere game and apply it to medical practice is irresponsible. The chief justification of scientific diagnosis is its applicability to medical treatment. Yet certain treatments, though based on tradition without any scientific basis, claim some measure of success (e.g., electroconvulsive therapy for depression), while other treatments, though scientifically justified, fail. This discrepancy indicates that scientific or intellectual success in the lab may not translate into scientifically attested technological success. This will be examined a little more in depth later on. Right now suffice it to show that there is a difference between explanation and application.

Most medical texts concerning technology tacitly suggest that
(1) all technology is rational,
or that
(2) all technology practiced in established, respectable institutions is rational,
or at the very least that
(3) rational people use only rational technology,
or, as the last resort, that
(4) rational people ought to confine their conduct to established rational practices.

Yet there are obvious exceptions to the claim (4) since the obligation it refers to is applicable without exception only in cases in which responsibility for the life and death of patients is at stake. (Since this is not articulated in the medical texts, it is easy to confuse these four different variants, and often enough even expert writers do so.) Little ingenuity is required to discover justifiable counter-examples to all four variants in current proper medical practice.

Since there is no consensus as to what rationality is, the situation is problematic. Yet enough is agreed about rationality to clarify differences between scientific and technological rationality that matter here. At the very least, scientific rationality embraces rational disagreement (as hopefully leading to crucial experiments). Technological rationality embraces consensus (since practices publicly recognized are social institutions). From a practical point of view, scientific explanation need no justification since science may be viewed as a pure game, whereas technological practices are required to have justification prior to their implementation, and the requirement is made and specified by law: its aim is to determine responsible behavior in technology. The simplest example in science is any disagreement among scientists, whether about quanta or about the etiology of cancer. A simple example in technology is the United States Federal Food and Drug Administration's decision to issue or withdraw a license for the use of any drug, from thalidomide to chloramphenicol. Responsibility is required on both personal and institutional levels. From the institutional perspective, there are legal standards for the responsible conduct in testing procedures before their implementation.

The consensus concerning practice, being an institution, does not invite conviction; on the contrary, the possibility of reform and improvement depends on the possibility of dissent. In some cases dissent about practice is justified by strong criticism and later by successful reform; in some cases it is rational even if misguided, and all that is required is either a criticism of the criticism and of the proposal for a reform or a test of the proposal; in some cases dissent is *a priori* not rational. This is why criticism is aired and discussed and assessed on all levels. For example, there is some sharp criticism of the use of electroconvulsive therapy; the criticism of this therapy cannot possibly be met without finding out first whether it is rational to begin with: this question naturally precedes the question whether it is valid. Even when a criticism is valid and the critics offer new alternatives to traditional practices, there is a need for consensus about it as expressed by a rather clear-cut change in the law or the custom.

Rational technology ought to be responsible, yet standards of responsibility can be changed--hopefully for the better. Hence, standards of rationality can be changed too. This holds even for pure science, as it is an abstraction of a concrete mode of human behavior. For example, treatment is to some extent always experimental and so it may partake in pure science. Thus, current standards of experimentation forbid trying a new concoction *in vivo* before trying it *in vitro*, and trying a new experiment on humans prior to trying it on animals. After the disaster of the births of deformed babies due to the use of thalidomide by pregnant women, the practice of offering new concoctions to pregnant women is no longer allowed prior to some tests or prior to warning.

Standard medical textbooks assume that technology should only be based on validated scientific knowledge. But there is no consensus as to what are the rules of scientific validation. Even if rules are clearly articulated, no technology can avoid guesses, and at times even bold guesses. In some cases, these guesses can be applied, which would be irresponsible, but not always so: for example, under emergency conditions, the refusal to try a new intervention could be irresponsible. (This is one of the many refutations of the view of rational therapeutic conduct presented above--of its four variants. The next paragraph offers more.)

Irrational practices are accepted in many areas of medicine. For example, many doctors conceal their past errors--because custom requires it. This custom is, of course irrational, especially since this way the errors are allowed to be repeated and their repetition is justified through incorrect statistics. The reason for the traditional concealment of errors is the irresponsible demand that physicians be utterly reliable. Even the demand to avoid all avoidable error is irrational, since it amounts to the requirement to check for each step for each possible medical error and this requirement reduces drastically the number of moves a physician can make.

Another example of irrational practice is the practice, imposed by laws in some countries, of expecting of all physicians to be available at all times. It may lead to physician's total collapse. Emergency rooms often overflow, and interns are often expected to be on call for thirty-six hours straight; in such cases practitioners are so tired that they are often in a daze. This can lead to gross errors. The explanation given for forcing tired practitioners to continue come what may, is that they have no replacement. Surely the more rational, and thus more responsible, option is to leave a station unattended rather than to keep it functioning in such a manner, and this option is always available. The situation here described has been observed in respectable institutions and these observations are testimony to irresponsible and irrational conduct even there. The claim that the situation is under study only enhances this point: prior to any study one can institute simple measures and controls to prevent excessive, namely gravely harmful, over-tiredness in medical practice-- though, of course, studies of such matters are indeed highly recommended, and even urgently so. To date no sufficient study has been conducted of the ill effects of the near-exhaustion of interns and residents in emergency rooms and wards. Research concerning the ubiquitous practices of excessive driving and night-shift work, indicates that tiredness dramatically reduces efficiency, especially on matters of vital decisions. This fact has induced tremendous reforms in practices of even the most demanding institutions (such as the Israeli Defence Force). Yet there remains immense pressure to increase the load on medical practitioners and to compensate this by other means. Against this tendency there is a distinct need to institute public controls. The example of limiting the monthly hours permitted for civil aviation pilots to sit in the cockpit is no different from the case at hand, yet tradition prevents rational legislation in the case of over-worked physicians.

The irrational practice of overworking interns, incidentally, is a major tool of discriminating against women in a society in which family burdens are not shared equally by the two adult partners. In such a society most women interns cannot afford to stay on the ward around the clock every day. This fact will hopefully call attention to the need to abolish this systematically imposed overwork which is harmful not only to women.

4. RESPONSIBLE AND IRRESPONSIBLE CONDUCT

How can irresponsible behavior be demarcated? The practice of medical diagnosis is guided not only by medical textbooks but also by the extant diagnostic custom, which is a social institution or a consensus, and which helps render diagnosis more responsible than were it guided by the medical textbook alone. To observe the difference between current medical textbook and custom one should return to the simplest possible case of responsible error. According to the medical textbook, this difference does not exist except in clearly and explicitly specified cases, such as the resection of an appendix with no appendicitis to justify the operation nonetheless. The medical textbook admits that twenty to twenty-five percent of all cases of appendectomy might be unnecessary yet are seen as responsible errors. (To consider an appendectomy without appendicitis not an error is to say that the surgeon removed a healthy appendix knowing that it is healthy. This, obviously, is flatly irresponsible.) How then is the standard of responsible error determined?

The medical textbook should answer this question: How is the standard of responsibility in diagnostic practice to be determined? Indeed, some examples of parts of this answer can be found there, or elsewhere in the professional literature. But unless some criterion is offered, there could hardly be a critical examination of assessments of current practices. Criteria of permissible error, even partial ones, may help spot imperfection more efficiently than mere spot-checking. To quite an extent, the medical textbook is written with an eye toward some limitations on reasonable error or permissible error, as prescribed by custom, tradition, and the law, just like the limitations on other technological activities. Nevertheless, in empirical fact the medical textbook is regularly in conflict with custom, tradition and the law, just as it is in conflict with technical and budgetary constraints. Interns are shocked and surprised at what they encounter in clinical practice. They must learn to leave the idealized medical textbook and learn to accept compromises imposed by the gap between the idealized and the real. This gap should be narrowed.

Systems theory and computer theory may help tackle some general, and often overlooked and systematically ignored questions, such as the question posed here: How is the standard of responsibility in diagnostic practice to be determined? Serious outlines of this question are nowhere to be found. Nevertheless, enough is known about customary research procedures to enable one to envisage simple starting points: studies of attitudes and statistical studies of reported errors are so prevalent that any expert in such matters may help any medical researcher study the primary facts of the matter. One may then easily check the worst divergences between practices and the law; one may then study the question: What is preferable, a better implementation of the law or legal reform? One may easily spot the most unacceptable yet legally permitted practices and study the way they may be reformed--whether administratively or legally. One may also devise the reform of the standard medical textbook, which is urgently required anyway. Suffice it here to observe that publishers of medical texts and medical associations and schools have an enormous influence on current practices, but regrettably they are seldom concerned with exposing recalcitrant traditions. Their chief concern is the presentation of some advanced diagnostic techniques in the abstract. They do not even mention the legally problematic cases. Most of the legally problematic cases end without dispute; alternatively, the disputes usually do not end up with litigation; and when litigation turns up, it is usually settled out of court. This way lawyers have a great influence on medical practices which is seldom aired for critical examination. Lawyers have to take account of the complexity of current practice and of the various factors involved, from legal to institutional to scientific; the current medical textbook ignores the complexities involved and offers idealizations which are too often quite untenable: the study of medical practice will greatly improve by the addition to it of some minimal legal and social considerations of permissible and impermissible error. It will thereby become less terrifying to students.

As to standards of responsibility (and thus of rationality, since, we remember, rationality and responsibility go together) it is most reasonable to expect standards to improve, since, at the very least we all agree that to commit any error which is both easily avoidable and possibly harmful is clearly irresponsible. The more is learned, the more is known as to how to avoid error. For example, early in the last century physicians were not seen as irresponsible when they unknowingly infected patients with childbirth fever. When Ignaz Semmelweis informed his colleagues that they were infecting patients and causing death in the maternity ward, things changed. Out of a sense of injury, the leaders of the profession dismissed him instead of instituting an examination of his claim. They thereby acted irresponsibly (and caused injury to the public at large, to the profession and above all to their patients). We should stress, however, that the damage discovered by Semmelweis was not necessarily irresponsible beforehand: it was done in an unsuspecting manner and it was done by the instituted mode of treatment.

The example cited is old. The harm discovered by Semmelweis well over a century ago has been rectified by succeeding generations, chiefly under the influence of Lister and Pasteur--also about a century ago. Some commentators suggest that such cases as this century-old example are no longer possible. They claim that such errors were part of traditional medicine, which medicine has since been replaced by scientific medicine in a radical manner, so that such errors have been eradicated for good. This claim is erroneous for the following four reasons.

Traditional medicine cannot be radically replaced by scientific medicine: only where a scientifically attested treatment replaces a traditional one is this at all possible; otherwise a traditional treatment can at most be abolished, not replaced. All treatment of incurable disease fall under this category as soon as their status as incurable (by current means) is scientifically attested. Second, a scientifically attested technique may be scientifically overturned. Thus, the techniques of Lister and of Pasteur have largely been replaced, and their replacements have also been replaced, yet hospitals still do infect. The infections in hospitals become irresponsible when left unstudied or unattended to, especially when scientific studies recommend change in the hospital's environment. Third, the demand that medicine be utterly error-free blocks information about error and is thus conducive to irresponsibility. The most weighty reason, however, is that there is no ideal treatment and scarcely a general theory of treatment available. Available, however, is what is institutionally or conventionally viewed as treatment in a given stage of the growth of medicine, and there has been some improvement in the institution or convention concerning treatment.

Examples. Surgery, which is persistently considered treatment, was repeatedly altered, and even the very concept of surgery was repeatedly altered (compare bloodletting and transfusion). Similarly, it is hard to justify the fact that self-administered ingestion of chemicals is treatment only when the drug is a prescribed drug and a qualified physician prescribes it, its dosage, and the duration of the self-administration. The harder question is: When should self-administration of a diet instead of a drug count as treatment? Today diets are self-administered as a rule, yet before the scientific revolution diets were staple treatment items. Later they dropped out from standard treatment, or at least became marginal, despite the fact that for centuries the only known cures for scurvy and for pellagra were diets of certain specifications. With the discovery of vitamins and of the value of certain minerals, diets once again became more central. But they moved to the domain of preventive medicine and public health, which are traditionally non-treatment oriented. This changed somewhat with the discovery of disease-specific diets, such as salt-free diets for high blood pressure and low-cholesterol diets for atherosclerosis. With the growing fashion of diets, whether as preventive measures, especially in cardiovascular diseases and cancer, or as attempts to stay healthy, attractive and young as long as possible, at times the role of the prescriber becomes increasingly problematic and quite irresponsibly so. Repeated reports were made of considerable specific damages which some diets have caused. The same holds for physical exercise, from jogging to calisthenics.

Hard core medical fields may also be problematic this way. The elasticity of the concept of surgery makes some sorts of surgery problematic as to whether they constitute treatment or not. For example, it is not clear what sidelines that go with surgery are integrative parts of it. Which pre-surgical and post-surgical tests should count as integral parts of surgery? Some examples, such as electrocardiographic readings, may count as parts of surgery. Certain tests and follow-ups, known by the blanket title of post-operative care, may but need not be viewed as parts of the operation. Post-operative care known as physiotherapy, nursing, and rehabilitation do not count as surgery. Even internal medical care eventuated by surgery does not always count as surgery. The distinction is not clear enough.

This absence of any explicit demarcation of treatment may be due to the absence of a general theory of treatment, and it may be due to some pressing need to delineate more clearly the boundaries of the field of medicine in general and its inter-relations with other fields. The mechanist view of disease and its surgical or quasi-surgical treatment is conducive to the evasion of medical responsibility for the "softer" aspects of medical care. Medical practitioners claim that those are not medical proper and therefore the responsibility of para-medical professionals. This renders the problem one of a division of labor. The problem comes up forcefully when psychiatric treatment is considered. The physicians' lack of background knowledge in psychophysiology, psychopathology and psychotherapy encourages the split between the "hard" and the "soft" aspects of treatment yet not without the high price of giving up the integrated view of the diagnosis and treatment. Economic, administrative and legal pressures (including the refusal of insurance companies to pay for post-operative rehabilitative care, for cosmetic operations or for psychotherapy, and the insistence of the law on viewing the physician as the bearer of primary responsibility for the overall treatment within the medical context), all these may demand such a demarcation of treatment from near-treatment. We may wish to base it on rational theoretical consideration as well as on pragmatic ones. A more systemic approach could invite a less dismissive attitude and offer an integrative framework within which all aspects--hard and soft--of diagnosis and treatment could be subjected to critical examination and subsequent evaluation.

These are all examples of the lack of a clear demarcation of treatment and with it the ambiguity of the broader concept of medical technology. Many factors prevent clarification. The most crass expression of this is seen in the focus and practice of medical insurance, where different underwriters use different definitions. Particularly, the boundary between diagnosis and treatment is necessarily vague and variable since prognosis is also a means of testing diagnosis and/or treatment. Moreover, it is most difficult to decide responsibly when to alter a method of diagnosis or treatment or to implement a new one. It is well known that a new method of diagnosis or treatment, even after it has passed all tests prior to its first implementation, is still rather risky. (In other words, the best test is in the field, *in vivo*). The availability of insurance money for a procedure offers an incentive to perform it. This clearly could depict a conflict of interests.

It is all too obvious yet seldom noticed that any conflict of interests concerning treatment of necessity spills over from treatment to diagnosis: when treatment is performed in the interest of the treating physician, then it is backed by a diagnosis performed in the same interest. If, for example, an unneeded cesarean section is performed, the ill-treatment is unavoidably endorsed by a misdiagnosis.

There is a need for criteria for responsible conduct in rational medical technology in general, and of rational diagnostic technology in particular. Yet it is important to insist neither on any clear-cut dividing lines nor on total risk avoidance. (Thus, though some cases clearly call for a cesarean section and some call for its avoidance, there is a broad grey area of a variety of situations in which the less risky option is a disputed matter.) Responsibility, therefore, is always a matter open to revision. Risk taking may be deemed calculated risk in one style of knowledge but not in another. Progress not only in medical theory but also in epidemiology or systems analysis may deprive a reasonable risk taking of its reasonableness, and thus make it irresponsible. It is easy to understand why medical practitioners may consider developments in extra-medical studies irrelevant to their tasks. Yet instances where such studies are relevant exist--this will be discussed later on--and the profession at large can be informed about it.

5. BETWEEN THEORY AND PRACTICE

No matter how medical science is viewed--as a mere game or as a practical technology--scientific diagnosis is its experimental sub-system and the diagnostician functioning as a researcher is

directed to search for experimental criticism of theories, for
what is variably known as the (scientific or not) rare case, odd
case, exception to the rule, or anomaly. This case is the one
that staunchly and repeatedly (repeatability is of cardinal
import in science) refuses to fit present theories and, thereby,
serves as an incentive to change our scientific views;   in
diagnostics it is to change our views concerning the very nature
of a disease. This consideration, too, is not new. Its *locus
classicus* in the literature concerning medical investigations is
Claude Bernard's classic *Introduction to the Study of
Experimental Medicine.* Later on the philosopher Karl Popper
offered the view of scientific method, here repeatedly exploited,
which is based on the idea of the importance of repeatable odd
occurrences for science at large. Potential odd cases or cases of
anomaly are important since they serve as reports of facts
whose admission amounts to the admission of falsehood of given
conjectures or hypotheses or theories, thus making them
testable and thus granting them empirical character. When a
conjecture is refuted, the road is opened for the search for a
better substitute. Also, when a conjecture or hypothesis or
theory is refuted, knowledge of the limits of its applicability is
thereby acquired.

Thus, diagnosis in the service of science has no
guidelines other than those which enhance and further curiosity:
a healthy critical attitude and the search for new knowledge.
Diagnosis in the service of science is impeded by the rules of
medical practice which enforce responsibility, chiefly towards
patients but also towards the public at large, and limits the
use of humans as guinea pigs to extremely rare and well
supervised cases. Diagnosis in science and diagnosis in medical
practice are likewise different in orientation, since normal
medical practice usually aims at the common patterns and not
at the single specific case, especially the odd one. This is a
crucial point: rules can apply only where there are generalities,
even though they are often meant to apply to single cases,
since the single cases to which the rule is to apply has to be
specified as belonging to some general set. But one never
knows for sure that a given case belongs to a given general set:
one is never sure that this patient has that disease. Such
determinations always involve decisions and thus raise the
question of responsibility, for which examples abound. Patient-

oriented diagnosis does not impose differentiation between cases treatable in the same manner;   science-oriented diagnosis is preferable, since it is concerned with patients more efficiently.

Quite generally, doubts concerning new treatments are of necessity diagnostic and often even differentially so: when a new cure is offered, the problem of the responsible way to handle it are often solved by monitoring the new cure and finding a new differential diagnosis.   For example, the failure of the cure of some cases of pneumonia by antibiotics led to the discovery of viral pneumonia.

Of course, progress takes time, and in the mean time problems may be both pressing and difficult--at least as long as there are unclear criteria for the applicability of innovations. Criteria are readily available for common patterns, but each innovation is partly uncommon.   And so the patient on whom a new method in medicine is used is always a guinea pig to some extent, who should preferably be a hopeless patient whose informed consent is very carefully procured.   At times this practice solves the problem, and it is viewed as satisfactory. But this way out may be too problematic because there may be great pressure to use the new treatment on the supposition that the risk of using it is deemed small and the expected benefit high.   This supposition is true for some cases.   In all cases the question arises: Is this supposition true in the case at hand?   This is not easy to decide: it is almost impossible to explicate the rationale of a new treatment without appropriate diagnosis, which is sometimes also new, especially in cases involving new treatments;   these treatments are offered only to some cases in a given general set, which can be decided by some new techniques of differential diagnosis, such as the techniques that differentiate bacterial and viral pneumonia.

The suggestion that the growth of knowledge strips scientifically-established responsible practices of their high status is scarcely surprising, and would have been taken for granted in medical circles,   were scientific medical knowledge not be deemed error-free.   After all, practices that only a decade ago were highly successful and deemed scientific *par excellence*, are no longer considered good enough, and their application is no longer responsible.   X-ray shadowgrams are the paradigm.

The influence of epidemiology on diagnosis--and hence on treatment--is a repeated point of the present study: as long as diagnosis is not complete, it is directed by epidemiological knowledge; and diagnosis is hardly ever complete. Yet this is not the end of the story: epidemiology and diagnosis may mislead each other. Indeed, a standard item of diagnostic responsibility in medical practice--one which may only be eliminated by the widespread use of a comprehensive computer-assisted medical diagnostic service--is the generation of epidemiological statistics by the way of diagnosis in the following customary way. Usually a physician diagnoses a case as that of a given disease, rightly or wrongly. Another physician learns about it and meets a patient with sufficiently similar symptoms. Almost inevitably, the information about a previous diagnosis increases, however slightly, the disposition to diagnose the new case as a case of the same disease. However slightly the disposition grows, a reiteration may lead very soon to a near-certainty by the inexorable laws of probability. This near-certainty may be of a correct diagnosis. If there are other prevalent diseases, sharing some symptoms with the one statistically reinforced, it may turn out to be an incorrect diagnosis. This is particularly obvious in cases of epidemics, where initial diagnoses are almost invariably reinforced--especially an epidemic of a new disease--and largely also because physicians work under mounting pressure. Public health officials, aware of this, will perform independent tests, both diagnostic and statistical. This reduces the error, but it does not fully eliminate it.

The final item to be discussed here is the usefulness of systems analysis for the reducing of the rate of tolerable error or of responsible error within medical practice. Better management of the medical system may improve its efficiency and therefore its rate of saving lives. Rather than consider specific management tools which may improve the performance of the diagnostic system, we may consider the management of error and risk taking in the system.

It would be easy to examine the situation, if there were universal criteria generally agreed upon. No such criteria are available, especially because of the widespread view that medicine is, or ought to be, quite scientific and that science is by definition error-free. Contrary to this there is the quite generally endorsed rule: there can be no general criteria of responsibility and of rationality, except that whatever partial criteria are accepted, they must be alterable, in a way that will promote the correction of some past errors. Past errors are at times those committed by individual practitioners and at other times publicly instituted opinions and rules. Hence there must be public records of past errors and of their eliminations, and of the degrees of responsibility or irresponsibility involved in their very introduction in the first place. Likewise, there must be public records about tests of innovations prior to the permission granted to their wide introduction into public practice. To this day there is no independent public control over the introduction of surgical technology into the market. This state of affairs has demanded its toll from patients. Vineberg's cardiac surgery is a relevant example. This operation antedated the currently widespread coronary bypass cardiac surgery and consisted of the implantation of a thoracic blood vessel into the failing heart, with the intent to enrich its blood supply and thus its function. This operation had been performed for a very long time without any valid empirical support for its usefulness. The failure to adduce such support dictated a search for new therapeutic techniques.

The ideas concerning the responsible regulation of error management and risk taking in the public domain are well-known, but mainly from newspapers and gossip; they are discussed in either medical or philosophical texts much less frequently and not more thoroughly.

6. NEW ROLES FOR THE MONITOR

The monitoring of error demands that the monitor sub-system of the diagnostic system should be independent or, at least, monitored by another independent (sub-)system. For, if the monitor will depend on the diagnostic system, it may repeat its errors. Now the monitoring of error always sounds to some people as an expression of mistrust. This response is not restricted to medical practice alone. Social science is now agreed that the way to build trust is not to overtax it, that monitoring is the best tool for the general increase of trust. Moreover, the introduction of the question of trust into the discussion tends to blur the distinction between a *bona fide* unavoidable error and sheer negligence, and, with all the good will in the world, we can never rule out sheer negligence. This is how mistrust is created which monitoring can prevent: when an error is detected or suspected, the monitor's record helps decide what kind of error it was. But, of course, the modern monitor, which is connected to a computer and can be programmed and re-programmed, can do much more than that, as it can detect and help eliminate errors of kinds previously not suspected and not catered for. Here are instances of the new roles of the monitor.

The monitor can follow up treatment, compare prognosis with the course of events, and report variations, such as the failure to find a gall stone or an inflamed appendix during surgery based on diagnoses specifying their presence. There is a fundamental difference in the ethics of these two examples. An operation for a non-existent gall stone is based on an unacceptable diagnosis: since a gall stone can disappear, final diagnosis before surgery is imperative. This is not necessarily the case in unnecessary appendectomy. The commonly accepted norm for unnecessary appendectomies is that if they do not constitute more than twenty to twenty-five percent of the operations performed, then there was no mis-diagnosis involved. Since mis-diagnosing an inflamed appendix as normal is very hazardous, diagnosticians are expected to bend over in the direction of the less hazardous error: better a false positive diagnosis than a false negative one.

The monitor has to report all these matters, of course, but there is more to it. There is the matter of improved norms and the need to keep up with them and to monitor the progress. It has been reported recently that the diagnosis of appendicitis with the aid of ultrasonography combined with traditional clinical means can reduce the rate of error to around eight percent, even though ultrasonograms alone will have the error rate of about twenty percent. Another report reduces the error rate to four percent.

The monitor has still more to do: the question has been raised as to whether new diagnostic tools are operative in the field as well as in the initial test situation, which involves experts, of course, and whether in all field situations the tools operate equally well under all field conditions. This can be best decided with the help of a wide-range monitoring system. Variations in the interpretations of the reading of new diagnostic (and other) tools are as pervasive as those of the assessments of more standard clinical diagnoses. This may provoke some diagnosticians to compete with the computer with some interesting results: there was a report that in competition with computerized diagnosis the rate of erroneous diagnoses of appendicitis was reduced to seven percent. This shows that the combined use of tools and of more traditional methods is always open to great variations, and that the level of sophistication of the monitor can therefore always be raised.

There still is one factor closely linked to error, and the computer can aid in reducing that factor too. It has been observed repeatedly by industrial psychologists that stress is the result of the demand to make quick vital decisions and that fatigue and stress contribute significantly to the level of unavoidable error at work. A comprehensive computer-assisted medical diagnostic service may aid diagnostic rationality by transforming the unreasonable pressure to avoid all error to the reasonable demand to act responsibly, by helping determine standards of admissible error, and by making it clear which diagnostic errors are easily avoidable with the aid of computers. Then, when errors are committed nonetheless, they would *prima facie* count as a violation of the demands for rationality and responsibility. With the aid of computers it is easier to study the levels of pressure which raise errors considerably. This will only be so, however, if the computer service itself be regularly monitored and reformed whenever required. Otherwise the very same service will be a temptingly useful tool for evading responsibility by the very appeal to the prevalence of certain errors (even ones that can be eliminated with ease).

Chapter 7

DIAGNOSTIC PRACTICE

1. DIAGNOSTIC METHOD AND THE NULL CASE

As tradition prescribes, the process of diagnosis moves from specific symptoms and signs progressively to increasingly encompassing sets: syndromes, illnesses, and (when available) etiologies. This process raises a practical problem: How does the diagnostician determine the proper move from one step to the next, from initial symptoms and signs to intermediary stages, and from them to the higher sets, and so on to the final diagnosis? Symptoms are grouped into syndromes and syndromes into diseases, but these groupings are at times clearly influenced by etiologies. Hence, the traditional proper step-by-step method cannot be followed in practice. Moreover, even the first step is problematic: What phenomenon may be properly considered a symptom?

Here is the traditional definition of symptoms (literally, whatever falls together): whatever patients complain about, or the complaints themselves is a symptom. This definition was soon broadened, as an afterthought, to include all of the patients' reports, such as self-diagnoses, however objectionable, as well as signs, namely, whatever diagnosticians observe or elicit from patients. Still as a mere afterthought, the set of symptoms was further extended to include the whole anamnesis (literally, reminiscence): histories of patients that usually originate from any source, including physicians' records, family information, and so on. This information is not always reliable, despite attempts to standardize anamnesis and to improve it both by the introduction of some standard procedure and by encouraging problem-orientation.

Some symptoms are very specific to certain diseases. They are called pathognomonic (*pathos* means suffering, disease, and *gnomonikos* means judge). A pathognomic symptom, then, is specifically distinctive of a specific disease. An example of a pathognomonic symptom is the lesion on the penis which is characteristic of syphilis, called *chancre durum.*

Other symptoms are not that specific, such as very high fever, but are nonetheless diagnostically very significant. Others are extremely non-specific symptoms, such as headache, fatigue and anemia. The presence or absence of symptoms may support a diagnosis, or may put it in question. This means that there is a probability of a certain symptom occurring or not occurring in a given syndrome. One might expect one symptom to be essential and another likely in the make-up of a given disease; but some cases refuse to comply with this smooth description, for example a case of the flu: none of its symptoms are essential. This will probably remain the case until the tracing of the diverse flu viruses will become so commonplace as to count as a part of the diagnosis or until a cure for flu will be common or nonspecific--or until the views of what constitutes diagnosis will change in a radical manner.

To complicate things further, certain phenomena may count as symptoms in one situation but not in another. An example of this is high blood pressure. When this is observed, the activity of the body at the time of the examination may decide whether the observed case of high blood pressure is a symptom. At times the situation requires further examination before such a question can be properly decided. Case histories of patients may be relevant to proper decisions, such as whether an event or a characteristic is essential or incidental to the complaint under examination. Many examples testify to this, including the well-known case of shock, which may be due to a mistaken repetition of all sorts of treatment (from injection to blood transfusion), and may be unrelated to the illness or the treatment as a whole. In brief, disease entities are determined by alternative sets of actual or potential symptoms (including outcomes of diagnostic tests that are considered normal parts of the diagnostic process).

Medical diagnosis handles alternative sets of reports, and correlates them with probability measures to sets of possible alternative diseases. Given a syndrome, a search is instituted for a test with a few possible outcomes, each of which, added to the given syndrome, would tip probabilities considerably toward one diagnosis or another. For each step in this process of determining a diagnosis, the probability of success may be improved by a test. In order to test, hypotheses have to be invoked and expectations based on them generated; only then can the test bring about some useful diagnostic information.

Once a syndrome is given, the diagnostic question is this: Which diagnostic hypotheses are plausible in the situation? This leads to a deeper question: Which initial list of hypotheses should be made available to the diagnostician for this process of narrowing down the list in the study of a given syndrome? Given that initial list, how should selection from it be effected? Supposing the process is stuck, should the initial list be revised? At times this is advisable. When exactly should revision be tried out? Rational diagnosis demands that diagnosticians neither waver too much between alternative sets of initial hypotheses nor cling dogmatically to any one of them. This demand is not operative, of course.

The diagnostic process includes repeated decisions concerning moves from one step to the next. These decisions are made by following certain prescribed guidelines. We are now going to consider the advantages of feeding such guidelines to a computer program. By way of illustration, we will consider first the null case, or a case in which the patient suffers no illness at all.

The null case is the case of a patient who is quite well and reasonably healthy. At times the reason for the diagnosis of a null case is medical or at least quasi-medical, as in the case of a check-up or when there is a reasonable suspicion of illness due to some epidemic or to physical exhaustion. At times the reason for the diagnosis of a null case is extra-medical, whether legal or economic, as an examination after an accident or upon a discharge from a hospital, or upon enlisting, or when subscribing to some health insurance plan. At times the reason for the diagnosis of a null case is personal, as when one requests a diagnosis as a medical defense of one's absenteeism or truancy as an expression of hypochondriasis.

The literature ignores the null case almost entirely, and for the reason that it is considered not problematic or at least medically unproblematic so that medical discussion may overlook it. Yet the discussion of diagnostics must take account of the null case, and for a few strong reasons.

Generally speaking, null cases appear regularly; they are unavoidable and can be quite problematic. They are even problematic within the field of diagnosis as a whole, even on the supposition that they are rather marginal. They are seldom conspicuous and seldom unproblematic even if the problems they pose are admittedly extra-medical. Also, as the boundaries of medicine are both vague and changing, one cannot declare all problems of all null cases clearly extra-medical. We shall see that hypochondriasis and malingering, in particular, are such null cases.

Even if every specific practical diagnostic study may be conducted while ignoring the null case, diagnostic theory, the study of diagnosis as such, cannot do so: the null case is more central to the general study of diagnostics as the framework of specific diagnostic studies. When diagnostics is formalized and follows explicit rules, including rules for follow-ups, then it must use statistics. Null cases are usually either absorbed into the sets of cases which they mimic or they drop out of statistics altogether. This leads to statistical errors and to errors of the diagnostic and other medical considerations based on statistics; it also leads to other diagnostic errors and to monitoring errors, usually either absorbed into the sets of cases which they mimic or they drop out of statistics altogether. This leads to statistical errors and to expensive diagnostic errors (such as untimely discharges from hospitals). The desire to avoid these errors and expenses is best met by explicit policies which should lead to the incorporation of null cases as null cases from the start; these policies have to be determined, tested and possibly applied.

The simplest null case possible is one diagnosed as conforming to the null hypothesis: the patient is healthy (or recovered) by accepted standards. Possibly no complaint was made and no sign was detected. Possibly the reported complaint is not a symptom or it is a symptom belonging to no syndrome and thus to no disease: it may be a lie, an error, or an expression of the patient's malingering.

Some null cases are not coupled with a null diagnosis. For example, mild harmless treatments are prescribed to patients who are not ill--perhaps under the patients' pressure, or because of mistakes, or for some extra medical reason. The result is over-treatment, however mild and harmless. Yet over-treatment can be harmful, especially since often and quite erroneously, over-cautious diagnosis and subsequent over-treatment are viewed as the opposite of negligence. It is well known that some healthy patients are over-treated with ill effects. These ill effects are iatrogenic diseases, which means diseases caused by medicine and/or physicians. These should not be confused with harmful or undesired side-effects caused by treatment and medication not considered redundant, or even the diseases caused by treatment which is not considered redundant. The iatrogenic disease always is, and side-effects usually are not, the outcome of utterly uncalled for treatments. As to uninvited damages caused by treatment, there are no differences between iatrogenic illness and side-effects *ex post*, but there is a great difference between them *ex ante* in that the ways to prevent them are different. And so, when the way to prevent a side-effect is the same as the way to prevent the iatrogenic disease, the side-effect in question is indeed considered iatrogenic; the most common example for this is the case of treatment instituted irresponsibly, as, for example, when the likelihood for the occurrence of side-effects is relatively high. Such conditions are called counter-indications and their existence should alert the prescribing physician to monitor the prescription in order to suggest a possible modification or a cancellation, perhaps in favor of non-treatment. The risks of counter-indications are weighed against the risks of no treatment. Hence, diagnosing a patient to be in worse condition than the facts warrant is dangerous. To avoid this danger a close study of the null case is required.

Even when correctly diagnosed, the null case may be problematic. Null patients invite no medical treatment, yet they are subject to public health considerations (to be discussed later on) and to medical treatment prescribed not on medical grounds, such as cosmetic surgery. (All surgery prescribed not for medical reasons has to be considered cosmetic: treatment is either functional or cosmetic.) The cosmetic surgeon must describe to the patron the significant possible side-effects or counter-indications, since the desired end-state is the guide of the operation. Any unexpected side-effect or counter-indications which surgeons prefer not to discuss in advance, is a sign of their irresponsibility. Nevertheless, not every performance of a cosmetic operation leading to side-effects is irresponsible, nor can all side-effects be eliminated.

Unfortunately, complication enters because it is quite hard to identify each null diagnostic encounter as one of the cases mentioned thus far, namely, the cases of a patient recognized as not ill; as not ill but treated due to some error; or as not ill and treated for non-medical (cosmetic) purposes. The complication is multiplied by lingering defunct diseases and fashionable diseases, those considered not diseases proper. These add to the list of possible null cases and should not be deleted from national health statistics on account of their not being real: they should be deleted only after they have dropped out of diagnostic practice. The complications which these cases invoke are multiplied in the situations in which even experts disagree on whether a given illness is real. Mental illness can serve as an example. Some view it as a medical illness proper, one that is here to say, as an illness of some unknown organic etiology. Others view it as a non-organic illness but still of medical concern. Still others, such as Thomas Szasz, see mental illness itself as a null disease, or, worse, as a iatrogenic disease, namely as damage to health caused by diagnosis.

To see how problematic null cases may be, let us consider the way in which diagnostic expert-systems software programs should handle them. Any attempt to program the system to help eliminate as many cases of null patients and prevent as much damage due to iatrogenic diseases as possible, regardless of the question of cost effectiveness, will effect a decisive deterioration of the overall efficiency of the service by making it too expensive. Such a program will have to alert diagnosticians repeatedly to all possible errors, including ones very rarely committed. Consequently, the service will be employed less frequently than that of optimal efficiency and both diagnosis and treatment will become more perfunctory. The overall effect of disregarding cost effectiveness is usually the reduction of efficiency, usually in an effort to increase it beyond capacity.

The guideline for any diagnostic expert-systems software is to minimize damage due to the null case and the cost of doing so; since these two guidelines work in opposite directions, one has to balance them optimally: the cost of reducing damage should always be limited by the cost of the damage itself. One can list the damages due to each kind of null case: the cost of unnecessary treatment to patient and to society, the damage due to the iatrogenic disease which an unnecessary treatment may cause, and so on. As to the damages caused by minimizing damage due to the presence of null cases, it is the reduced cost effectiveness of the expert-systems software, its resultant under-utilization, the distrust of the public which it may cause, the fear that the vigilant search for null cases will lead to under-treatment, and so on. To assess each of these damages, their frequency within the population has to be determined if and when it can be done.

Nevertheless, since most diagnostic encounters end up with no prescription, it may be surprising that most expert-system programs ignore the complex problems the null illness raises. Null cases are most often treated intuitively. When they are deemed illnesses and no treatment is prescribed, there is perhaps no harm done. But sooner or later the oversight of null cases by programmers may lead to some harmful results. The case of computer-assisted diagnosis resembles that of pharmacotherapy. As computer expert-system software programs become more powerful and widespread, their side-effects will become apparent. This might be so especially with programs geared to self-diagnosis in emergency conditions, which is the case with heart problem programs now in service.

These problems are not specific to medicine, but to services in general. Services are generally perfunctory in impoverished populations; in affluent populations they tend to be excessive and thus possibly harmful. Both kinds of population usually suffer a low level of critical ability to assess effectively the required level of the service. Here the low cost or cost benefit of the comprehensive computer-assisted medical diagnostic service is essential: cost effectiveness may justify the reprogramming of the computer in an effort to meet some specific iatrogenic problems, but not all of them. As the service improves so might be its ability to handle such problems, and these problems cannot even be stated without explicit and statistical reference to null cases.

The same computer service may offer to customers concerned with some significant issues alternative lists of possible diagnoses for comparison. It is often easier and cheaper to consult a computer than a consultant, especially when the consultation is about the desirability of (costly) consultation. For example, books written for the general public are largely used for initial self-diagnoses. To be more useful, the lists of possible diagnoses have to refer to practicable treatments and their possible side-effects. Yet the parts of the computer service devoted to lists of possible side-effects (of diagnosis and/or of treatment) must remain peripheral, so as not to clutter the system and render it too costly. Treating side-effects must remain a sideline, since every diagnosis and treatment can and often does have some side-effects.

It looks trivial to propose that side-effects must stay in the sidelines and perhaps it is. The fact remains that the research into applied medicine takes this proposal as extremely hard to implement. One meets complaints that certain treatments and drugs are out of the market or out of the over-the-counter market for poor reasons, and *vice versa*. The more the details of any such a complaint are discussed, the likelier it is that the discussion will lead to this question: what criteria does one use as guidelines for practice? There are specific criteria that differ from one case to another--at times considerably so. If the criteria themselves are questioned, then the debate may be marooned, and it can only get afloat again when it is noted that what keeps a side-effect (or anything else, for that matter) on the sidelines is its low cost effectiveness. This, too, may be a triviality, yet spelling out its chief significant consequences is a somewhat complicated exercise, here left to the reader. When doing this exercise, the reader may wonder about calculating cost effectiveness. For this an even more trivial guideline may prove helpful: when in difficulty, revert to the principle that any treatment that is worse than the ill it comes to cure should be overruled; in the present study the concern is with diagnosis only, but monitoring of treatment, including the continuous assessment of its cost, is here included under the rubric of diagnosis. The monitor can say, and at times should say, the initial diagnosis was too severe, or the cure is becoming worse than the ill it comes to cure, or the effectiveness of the treatment is becoming so minute as to be considered redundant.

(Here the unavoidable link between epidemiology and prescription is strongest: all considerations of side-effects, of the cost of an error, and so on, are statistical. The treatment or the medication with intolerable side-effects drops out of the market, at times due to the initiative of public health offices, food-and-drug administrations, and consumer unions. At times complaints are levelled against too lax or too drastic measures. This is inevitable. Some of the complaints are obviously valid and are not taken up because of some sociocultural factors and because of strong lobbies whose activities may cost the nation in an estimated annual rate of avoidable deaths. A comprehensive computer-assisted medical diagnostic service would be an excellent tool for implementing a more rational attitude to public control of side-effects of licensed treatments and medications.)

What has been said thus far does not exhaust the discussion of the null case. The matter will be met again in the discussion below of the most difficult null case--that of malingering. The null case was presented only to help introduce the discussion of the rational diagnostic method in general (especially since dependence of diagnosis on statistics is stressed together with the systematic errors in current statistics due to the oversight of null cases).

2. RATIONAL DIAGNOSTIC COMPUTATIONS

The rational diagnostic method in general terms is the effective correlation of illness to syndrome, by making diagnosis differential whenever possible. Differential diagnosis requires limiting the diagnostic information to minimum while seeking the additional, differential item--the one that can serve as a crucial test between given competing diagnostic hypotheses. The most expensive diagnostic practice is the jumble of data kept by physicians and by medical institutions about each passerby. These assorted facts are kept with the hope that they fall into place in a natural system of possible diseases from which diagnosticians will pick one day the proper one to guide some similar case. This hope is not based on facts: the random and unsystematic records peter out. Before that they often clutter diagnostic pictures. The tradition of keeping full records is costly, and even when useful it is inefficient. With the advent of computer technology and the availability of cheap vast memory banks, the traditional custom of indiscriminately gathering all sorts of data may get a new lease on life and clutter computer memory banks to the point of overflow.

Rather than alter their viewpoint, the inductivist believers in heaps of data invented a new computer technology called data retrieval. Though data retrieval techniques are at times useful, they were never meant to help make sense of any data. The comprehensive computer-assisted medical diagnostic service cannot absorb data indiscriminately, and, if forced, will soon also indiscriminately return such data as its natural overflow. Data retrieval technologists already handle data overflow by consigning probably useless data to increasingly remote parts of computer memory, where they can unobtrusively get lost. Such work invites cost-effectiveness considerations. These are made on the basis of some discriminations. Discrimination means the use of given sets of hypotheses to decide relevance, and with the growth of knowledge these sets change. To allow reassessment and rational change, computer programs need flexibility functions. These functions, however, are utterly irrelevant to programs which are indifferent to change. Of necessity, programs which jumble all data, relevant and irrelevant alike, make the whole system hopelessly confused, overstuffed and obsolescent. Data retrieval experts often act accordingly and dump redundant information. Yet they may do so intuitively, and the need to do so in an explicit and formal fashion is becoming increasingly apparent. The formal tools, such as queuing theory, cost-effectiveness theory and (linear) programming are available to some extent and at times they are used.

Translated to diagnostics, this discussion invites the study of the null cases. Since all effective diagnosis is differential, and hence statistical, the proper mode of reasoning of even simple problems is too involved even for excellent diagnosticians, not to mention ordinary ones, whose knowledge of statistics, queuing theory and cost-effectiveness computations is limited, and whose computational capacities are low. Rather than acquire these essential skills they use intuition which totally mixes null cases with mild ones. Though intuition is always indispensable, whenever replaceable by tested theory and/or computational devices, it should give way so as to be free to do what computers cannot do. Statistical computation as well as some aspects of medical intuition are programmable and thus may influence each other in an adequate computer-assisted diagnostic system. Particularly when statistics is used for follow-ups, the statistics must account for null cases as a separate category; particularly so since the patient discharged later than necessary is a null case. Computers make no decisions; nor can they replace human intuition; when they look as if they do, then they only represent some programmers, for better or for worse. It thus remains for some person in some responsible position to judge what is going on. The case of responsibility for decisions falling between programmer and user is so natural that it must be prevented by strong monitors, computerized as well as human. With normal programming, each diagnostic act should be delineated as an integrated system with the diagnostician in control, taking responsibility towards the patient. If the program encourages the oversight of null cases, then the results may become unpleasant in a short time.

Even simple diagnostic computations cannot be worked by most physicians, and even with the aid of computers one usually obtains poor results. Consider Bayes' theorem which permits, under the stringent condition of completeness, to conclude the probability of causes, given some effects, from all the probabilities of all possible effects given the complete list of possible causes. One may conclude the probability of the presence of a disease given a symptom, or the probability of an etiology given a syndrome, from the full set of probabilities of all syndromes given a full set of etiologies. Given lists of causes, effects, and probability for each cause that it leads to every given effect, and given that these three lists are complete, then the theorem tells us how to compute the probable cause of a given effect. Given a syndrome, there is a probability that it is caused by one etiology and another probability that it is caused by another, and so on. It may be hard to tell what is the complete list of causes and/or of effects and how they link with probabilities (except that by definition the probability of one given effect, given the disjunction of all of its possible causes, must amount to certainty). Given the three complete lists, however, then conversion is possible of the information of the probability of an effect given a cause, to the information of the probability of a cause given an effect. The computation is simple yet also bothersome and it usually cannot suffer shortcuts. Computers calculate this conversion in no time; computer diagnostic services should offer the background information required for this computation--or even the finished product of the calculation--provided these are considered necessary for a certain diagnostic procedure and provided the information in question is periodically revised. **It is impossible to apply Bayes' theorem without contradiction unless one is careful to include or to exclude systematically all null cases.** The latter option, excluding all null cases, is quite troublesome due to the very high frequency of diagnosis leading to no treatment.

The collectors of facts without discrimination have faith that facts fall into patterns when simple formal rules--simple algorithms, as the jargon goes--are applied to them, called the rules of induction, and then the collected facts will display the law of nature which have given birth to them. When pressed, these believers in the indiscriminate collection of facts invoke Bayes' theorem as the tool with which effects may be converted to their causes (thus eliciting etiologies from syndromes, for example). The inductivist believers, who practice the indiscriminating collection of facts, should have observed the following hard facts: (1) the collections of facts are useless in most cases, and (2) only further hypotheses grants some of them some significance or relevance. Instead of answering these empirical objections to their views, the collectors of facts without discrimination invoke a sophisticated field of computer technology--computer simulation, Artificial Intelligence or expert systems, declaring them able of emulating the expert diagnostician, as if this absolves them of the need to answer the objections. One may try to explain why inductivists assume that if a machine can emulate humans, then they need not answer these objections.

The method of creating Artificial Intelligence systems able to simulate human conduct is that of making hypotheses about the strategies endorsed by human beings under specific conditions. Were Artificial Intelligence systems constructed, they would be able to conduct research successfully. They would thus embody strategies for performing research successfully. These strategies will have to be algorithms, and such algorithms will be, by definition, the canons of the inductive method for which the inductivists always yearn. The computer-simulation students then are trying to construct hypotheses which, if successful, will make all hypotheses superfluous. What these Artificial Intelligence researchers do is, indeed, not too different from what the naive inductivists describe: they conjecture and formalize statements about the strategies employed by research scientists and then they feed them to computers: what they do when they study any human activity which they try to emulate is seek human behavior patterns--of chess playing, of diagnosing or of research. The idea of discovering algorithms for invention is of course paradoxical. To have total faith in it is to deny that invention is at all possible. Having no faith in it, however, does not require a denial that it can have some relative (never full) success, and then the success of simulations is instructive. In particular, any successful simulation suggests means of severe testing of hypotheses about current strategies and about their adequacy. The explicit description of current strategies and their successful emulation help put strategies under critical examination and possibly improving upon them or at least increasing the speed and efficiency of their implementations. The most powerful advantage of successful simulations, however, is that the boundary of their implementations can be studied: the partly successful program fails, it "cracks", when some unspecified context conditions have been violated. This way the limits of a program can be discovered and explicitly described; the new description will be of the context of the program's applicability.

The method of computer simulation is extremely limited. One limitation is technical: hypotheses about strategies at times may be tested, refuted and modified much more cheaply and quickly in more traditional ways. Another technical limitation of computer simulation is that it applies successfully only to highly specialized cases, only when the hypotheses about the cases and strategies are formalized, only after these were modified and withstood tests, and only where the broader contexts in which they function and are tested are relatively stable. The real defect of computer simulation as practiced today is not that it absorbs all the defects of the intelligent person it emulates but that it does so rigidly (it is "brittle") and perhaps irreparably: the only way to improve upon the costly and complex exercise of a given computer simulation is the re-doing of the whole exercise all over again.

The method of repeatedly starting more-or-less afresh is highly impractical--except in research. Here lies a tremendous optical illusion. For obvious reasons, enormous sums are invested in Artificial Intelligence research, in computer simulation, and in devising diagnostic expert-systems software programs. Financial interests are tremendous, and interested parties are prone to display their most successful outcomes. Rough cost-effectiveness computations suffice to illustrate the preferability of a comprehensive computer-assisted diagnostic service over the proliferation of specialized isolated diagnostic expert-systems services, especially since the former can easily absorb the useful fruits of the latter but not *vice versa*. This has become common knowledge among experts developing diverse commercial expert-systems software programs.

Analyses of computer programs are now a matter of a new, vast and important specialty. It turns out that even in purely mathematical calculations there exist programs too cumbersome for computation, yet easily replaceable with human-assisted programs which help obtain the required results. Since this is true even in purely mathematical matters, it holds *a fortiori* for human affairs--medical diagnosis in particular.

3. DIAGNOSIS AS PATTERN RECOGNITION

To view the diagnostic process realistically, it is advisable to start with its end: to generate an efficient and reliable link between symptoms and a diagnostic outcome, preferably a disease entity and still more preferably an etiology as well, and most preferably the best possible recommended treatment thrown in and a prognosis as well. The diagnostic process may involve questions and examinations. How much of all of this can be standardized, formalized and fed into computers?

This general question is better approached generally and abstractly. One may easily envisage a standardized list of symptoms, a standardized list of signs, all possible groupings of these, all possible syndromes and a probability function leading from each set of two groupings, of symptoms and signs, to syndromes or disease entities or etiologies. The mapping is called a pattern recognition. The most general form of pattern recognition is too vast to be of any use. At each stage, even the initial stage, diagnosis may be declared complete and the outcome a disease entity. The word "syndrome" then can be omitted altogether since it only indicates that the process of transition has already begun and is not yet declared complete. The result is a mapping of sets of symptoms onto sets of diagnostic outcomes which may include sets of symptoms, disease entities and etiologies. The set of initial diagnoses is a small subset of this mapping, where sets of symptoms are mapped into sets of symptoms. A more developed set of initial diagnoses is one which is obtained when all impossible mappings are eliminated, where impossibility is determined by the medical theory extant. The ideal final diagnosis maps each set of symptoms to one disease entity or, when available, to one etiology. This ideal cannot be written up formally. Therefore, to make the mapping of a pattern recognition more useful than the initial diagnostic mapping, and less useful than the ideal, some concrete elements may be added by applying certain computer-simulation techniques--of the kind already available in the diagnostic expert-systems literature which serves the computerized partial diagnostic services now on sale in the market. For example, the computer may list standard tests: all patients complaining of a given symptom showing a given sign must undergo urine analysis, or some specified X-ray examination, and so on. This narrows down the set of options available, but not far enough.

The next step is sophisticated and invites a revision of the little already said of the abstract idea of pattern recognition. So far each symptom and sign was listed, and all of their combinations were allowed (as syndromes). When too many such sets (syndromes) are available, a test may be used as a filter selecting from among them. This is a complex process of transition from one set of symptoms and signs to the next, with the omission of some items as not relevant but as incidental or coincidental and the addition of items elicited by questioning, further examination and tests. It is here that Kahneman and Tversky find a lot of intuitively acceptable gross errors: when the result of each step of the deliberation is finalized, the reasoning leading to it is overlooked; it is then taken as given when the next step of deliberation is undertaken. The overall line of reasoning is then illogical even if each step is logical and reasonable. Intuition permits this if the overall overview is not given to critical examination. Therefore, diagnosis must also include correction of errors. To that end it must include the process of checking final overall outcomes: after a series of transitions is completed and recommendation to view the outcome of the process as final is present, checking must take place, allowing for alternative corrections of possible errors. Intuitively speaking, pattern recognition moves back-and-forth--approaching the target and alternatively receding from it, so-to-speak.

The need to move back-and-forth was well-known even before Kahneman and Tversky offered it such strong rationale. For example, initial diagnosis may place the heart as the source of a problem, especially with a patient known to have a history of heart disease, even if the source is the alimentary tract. This is why a diagnostic process (including even traditional diagnosis) has to go back-and-forth, as it often does. How?

This question has already been answered: the choice of context prescribes to a large extent the way a symptom is diagnosed and finally treated. (The context in the above example was a sick heart, yet it has been altered and the new alternative context looked better suitable to the facts and a differential diagnosis than has favored it!) The set of alternative contexts--the para-text--is given and examined interactively. This is how things are done, and this is both the success and the limitation ("brittleness") of the existing diagnostic expert-systems software now on the market.

The diagnostic process must go back-and-forth in order to correct errors: an item may be erroneously misplaced and thus drop out of sight or omitted as irrelevant. In the general theory of pattern recognition this is known as noise: noise is every signal that is better ignored, even though its source may be a signal rather than the channel. Hence, viewing any signal noise is but a hypothesis, held tentatively till the whole message is deemed satisfactorily deciphered. As long as other signals are not finally deciphered, a signal omitted as noise is omitted only tentatively. The basic assumption of information theory is that noise is unavoidable and may be subdued by the use of redundant, i.e., probable, items of information. Hence, competing hypotheses are ever present, one will take one part of the given symptoms as noise, another will do so differently. Hence, comparing different final possible readings is a part of the process, and it is helped usually by checking details with patients. This testing is imperative, as an integral part of the test situation.

What makes pattern recognition so attractive and so amenable to computer techniques is precisely its flexibility. The flexibility is made indispensable as the end-product is selected *a priori* from a given set--here a set of patterns or disease entities--and tested for acceptability. Moreover, computers can emulate a human image-examination for given patterns, such as finding a picture of a tumor in a complex picture. It should be noticed that this advantage is limited to instances which are obvious except for their complexity, which is what invites the computer's service. For, what is the use of a computer's eye when it is less sharp and less reliable than a human eye? Answer: it can scan many images, omit the obviously wrong candidates (as construed by programmers) and leave the rest for a human to further examine so as to narrow down the possible options even further. The computer's work is always preliminary drudgery (speed and efficiency enable computers to perform unbelievable amounts of drudgery); the decision process is always human.

Finally, the role of filters should be noted. Consider a symptom a which is filterable if the finally diagnosed disease is b, but not to be filtered if the finally diagnosed disease is c. Example:

$$a = \text{headache}; \quad b = \text{flu}; \quad c = \text{meningitis}$$

The completion of the diagnostic process may be the decision in favor of option a, b, c, a combination of these three options or something else altogether. Only epidemiological statistics can help decide rationally the proper order of taking up these options for further examination. True, a patient's complaint is unique and statistics treats the unique part as sheer noise--at times with no justification at all. But since an approximation is a matter of cost effectiveness and is therefore unavoidable in all societies, except perhaps ones totally devoted to medicine. Once cost effectiveness is included, so is statistics, and once statistics is included, then making it cost effective is license for the introduction of the computer as a powerful tool for statistical calculations. The use of pattern recognition is the natural conclusion of this process.

Unavoidably, then, choice both precedes and influences the process of pattern recognition and even its end products. It is chiefly made by limiting the process *a priori* to a relatively small sets of patterns from which to choose. Therefore it is imperative that the choice should remain tentative, in the hope to have each item replaced with its improved variant--not necessarily from the initial partial set. Cost-effectiveness considerations also enter, and declare that all patterns which invite the same treatment should be deemed identical in the sense that differential diagnosis between them is a waste. Hence, also, when cost of treatment is too high, it is time to stop diagnosis. By the same token, when a complaint is not severe, optimal costs should be set low and raised if need be.

Chapter 8

SOME INTERFACES

OF MEDICAL DIAGNOSIS

1. INTRODUCTION

The clinical diagnostic encounter is an intimate affair subject to the rules of confidentiality. This fact makes everything directly related to diagnosis rather conservative in outlook, despite the effort of most diagnosticians to make use of the latest innovations in order to be of better assistance to their patients. This outcome creates inevitable tensions. The tensions are usually resolved as diagnosticians often tend to see themselves as funnels and sieves, with the funnel being as receptive to innovation as possible, and with the sieve being as conservative as required. This solution is not very happy, especially since it enhances the already strong disposition of physicians to be paternalistic. In its turn this disposition increases pressure on physicians and also tension between them and their patients, especially patients sensitive to paternalism. Here is the place for control over diagnosis--in single cases as well as in public health.

Nevertheless, the image of the physician as a sort of funnel-and-sieve is useful, especially when systematically considering what the funnel may be receptive to, when viewing diagnosis as having an interface with other services, in the systems approach sense, which permits not only interface, a boundary between two domains where exchange takes place, but also a joint domain or a significant overlap. As to the sieve, we are skeptical about physicians' ability to control well situations in which they are personally involved: we advocate a simpler and more powerful system of control as a public

- 155 -

monitoring system under democratic control. But we wish to deal with the funnel before we come to the sieve, the parts of the environment that provide much material that diagnosticians collect before sifting it.

The theory of diagnosis as pattern recognition, presented in the last chapter, makes it clear that diagnostics has a significant overlap with formal information theory. The overlap is not in that diagnostics is concerned with knowledge (gnosis = knowledge) and thus with information, since information theory is not concerned with information but with the free flow of the signals that carry it. The theory of diagnosis as pattern recognition suggests that there is an active interface between diagnostic theory and information theory. We will discuss this, theoretical aspect of information flow before coming to the interface between the clinical diagnostic system and two segments of the public control systems, the public-health system and the authorities issuing and revoking licenses.

2. INTERFACE OF INFORMATION THEORY AND DIAGNOSTICS

As long as the computer performs simple tedious jobs which involve practically no judgment--calculating, counting, analyzing chemicals, averaging sizes, and so on--overlooking its role as a diagnostic tool is usually permissible, and one may consider its introduction into the field of medical diagnosis not more problematic than the introduction of the high resolution optical microscope. The introduction of the electron microscope is different: unlike the optical microscope, the electron microscope offers a superposition of many layers, which must be separated, and which are separated by computer-assisted image-analysis techniques. Such techniques involve sophisticated judgments which are often fed into computer software wholesale, to go with the electron microscope and its associated hardware. The separation techniques are often tested before implementation--by quite independent means, such as x-ray diffraction. Yet once etched into the hardware or to the staple software computer programs, the errors they contain are fixed and often extremely hard to detect. The errors may be marginal--small and rare-- but this is no consolation to their victims. The simplest computer image enhancement techniques include techniques of different sorts for simple noise removal, excess shading removal, edge finding, symmetry image analyses, computer-assisted measurements, computer-assisted scene analyses, and more. These enter diagnostics by the extensive use made of them; they appear first in research, leading to new diagnostic and therapeutic techniques; then in special therapeutic systems; and finally as standard diagnostic tools. Examples: computerized axial tomography (CAT scan) and ultrasound.

Ever since the advent of scientific medicine in the last century, experts provide their inexpert colleagues certain techniques, without offering with them the thorough grounding in the relevant background theory. This resulted in very imperfect practices, and invited as partial compensation expert consultations in all unusual cases. It is important to notice this: much of the computer techniques which is used by average physicians who lack the expertise in nosology or in computer technology, are based on certain expert judgments which are not sufficiently familiar to them. In such cases, there is room for consultation with experts in nosology or in computer technology, especially since the existing diagnostic expert-systems software programs already in use may improve as they strongly interact with their users. This improvement may be a further screen which conceals errors of the initial design of the systems. As there are many strong competing computer techniques available for almost any service computers can offer the diagnostician, it is advisable to use competing systems simultaneously whenever this proves acceptable in terms of cost effectiveness, plus an obligatory (computer-assisted) alarm hooked to treatment monitor systems.

The general manner of handling computers in diagnosis is too complex in operational detail to be discussed here. It is mentioned as an item never to be overlooked though not on the agenda here. The mere piecemeal introduction of computer-assisted diagnostic technology cannot reduce all risks involved. (Remember Kahneman and Tversky: a set of decisions, each of which is consistent, need not constitute a consistent decision; even without involving conflicting decisions, the overall outcome may be bizarre.) For one thing, the samples handled by partial computer-assisted systems are not representative, and the growth of their application alters the nature of the ensemble using it in diverse uncontrolled ways.

In addition, the most successful techniques will be introduced first, and they will be very powerful and very appealing to the intuition of the expert diagnosticians. These will easily seduce them to overlook the judgment of the diagnostician and computer technologist who together had created the means for improving diagnostic techniques. A powerful example is the means to visualize non-visual sets, such as sound patterns, sets of distributions of (diagnostic) patterns (akin to weather charts), and (statistical) charts of alternative courses of given disease entities (which do not exist in the normal sense: the infecting agent, for instance, is more real than the disease it transmits). All of these new programs involve much judgment and yet are seductive since they are made to look intuitive. The simplest example of the danger this seductiveness involves is the noise removal which makes a picture so much clearer than it initially was; it makes a picture useless when taken as a means to detect noise! This is the standard problem with EEG deciphering. Computer technologists also operate intuitively and judge the outcome of their operations in some unproblematic intuitive ways (as in the operation of removing noise and thereby discovering the sought-for patterns--especially if the operation is the fiddling with dials). The resultant operations are both excellent and risky. Unless it passes critical examination, what seems intuitively useful and seductively unproblematic, may conceal objectionable errors. (Even an enhanced sahdowgram of a double pneumonia may be risky, as the two lungs look alike and so seem healthy.)

One kind of dangerous error necessarily involved in noise removal is the removal of all rare cases. It is cheaper by far to ignore patients suffering simultaneously from two unrelated diseases, for example, than to program for their inclusion. Consequently, the hope of such patients to attain a quick diagnosis--already very low as it is--will severely diminish with the increased employment of computer-assisted diagnostic techniques. Patients suffering from two connected diseases may have their diagnosis improved as a result of the introduction of computerassisted diagnosis, but this is small consolation for those suffering from two unrelated diseases. Every human system, and so every diagnostic system, has its victims. When young practitioners tend to perform differential diagnosis, their older peers discourage them from considering too soon the possibility of a rare disease. This is a matter of cost effectiveness; and the case of a dual disease with unrelated components is but a case of a rare disease. (The probability of two unrelated diseases is the product of their probabilities. Thus, if one in one hundred patients has a given disease, it is not a very rare case; but having two such diseases has the probability of one in ten thousand patients, which is rare enough to be ignored in the first instance.)

The program should include a constant reminder that it ignores rare diseases, as well as a strongly interactive routine to put a rare disease fast into the set of live options once it is reported in the vicinity, and which omits it again under some given conditions to be discussed and decided independently. A monitor may also be useful here--when applied to the system as a whole. Assuming that experts' performance is today poor in some sense, a computer may outdo all experts even while performing poorly in that same sense. This way the considerations aired here may still be premature. For diagnostic practices that are improved and for those that will improve soon, these matters get to the top of the agenda.

Rita C. Manning observes, in her "Why Sherlock Holmes Can't Be Replaced by an Expert System?", that an expert system cannot pose a problem well, nor can it discern the value of some data as relevant and of other data as confirmations, when these are unsuspected. So far she is right, of course, but then she also refers to flashes of genius; it is wonderful that some of us have access to flashes of genius, but this access should not be expected on a regular basis, not even from experts. Manning sounds as if she expects from experts but not from mere specialists to produce flashes of genius on a regular basis. If so, she expects too much from mere humans. Admittedly, there is a difference between expert and mere specialist, and admittedly this difference lies in the expert's superior performance on a regular basis; this superior performance, however, is not magical: it is due to the expert's wider knowledge by comparison to the mere specialist. The expert trains fewer experts than specialists, because the expert's deep knowledge takes a longer time to master than the recommendations based on it. Since expert systems replace experts in the capacity of givers of advice to specialists, not in the capacity of instructors, they resemble at best expert-assisted specialists, not experts: they represent no more than what an expert can teach any specialist: the application of a theory, and even the rudiments of the theory applied, not the familiarity with the theory required for the better comprehension and critical examination and assessment of the intricate aspects of tricky applications.

An illusion masks the fact that expert systems represent not experts but the average professionals who have learned from them. The illusion is due to the oversight of progress: the part of expert knowledge described by rules of procedure which thereby leaves the domain of personal expert knowledge and enters the public domain. Considering the very first expert system ever--of the mid-sixties and related to the spectral analysis of organic molecules--Pamela McCorduck states that it was "a complicated procedure", from "the domain of Ph.D. chemists". This is true but misleading: a decade or two earlier that very same procedure was the domain of Nobel laureates. If the program can be compared to Ph.D. chemists, then by the same token it can be compared to Nobel laureates. In a sense this is true: Louis Pasteur's discoveries, made with great ingenuity and effort, are common knowledge today. Expert systems include common knowledge, knowledge they render common and knowledge still limited to professionals. The expert knowledge which is available only to very select publics, is as yet not given to programming for computers.

The remarks made here are not meant as solutions to problems; they are meant to draw attention to problems systematically overlooked and to the need for solutions presented as public policies, publicly implemented and monitored. To take but one example, consider the need to place a rare disease quickly on the list of more likely candidates at the time it is reported. This may open the door to another undesirable practice: diagnosticians may easily be swayed by reports of infectious diseases, especially wild ones, to diagnose their patients as instances of a spreading epidemic and thus create a false impression. Usually such cases are mild and passing, except when they are corroborated. Nevertheless, it is not easy to draw the line between the undesirable tendency to overlook a rare case and the undesirable tendency to keep it constantly in mind. No optimal solution is available. But it is possible and desirable to put limits on what solutions are acceptable, and these limits should be observed; simple alarm systems could be developed to alert us to reconsider solutions we have decided to reject as too poor. Another example is the already discussed confusion regarding the use of the diagnostic expert-systems software programs concerning consultations with experts. Experts consultation are still important, and when advanced diagnostic techniques are used, engineers should also be available for consultation, rather than encourage users to lean too heavily on instruments.

3. INTERFACE OF PUBLIC HEALTH AND DIAGNOSTICS

Public health is of little interest to individual citizens these days except when certain epidemics make headlines or when general demographic factors connected with public health attract public attention. Yet this is an oversight: in modern industrial societies check-ups quite commonly concern most citizens personally, and the standards of check-ups as well as the details of their procedures are matters deeply dependent on public health. In both cases, their effectiveness and quality can be easily improved if a comprehensive computer-assisted medical diagnostic service be instituted. This is not to say that epidemics and check-ups are the only items which make public health a matter for concern and the comprehensive computer-assisted medical diagnostic service useful. Epidemics and check-ups are but the obvious extreme cases: once cost-effectiveness considerations are made explicit, public health statistics enter significantly and regularly into diagnostics. Hence, public health is of concern to the average diagnostician, though differently in affluent societies than in poor ones. Since public health is always a major factor in determining the order of priorities in matters of health, private and public, it is better to have the considerations and order of priorities explicit in order to open them to public scrutiny, that is to say, to bring them under democratic control.

Public control mechanisms are known primarily to be the institutional means for checking social action. The control measures concerning the clinic are no exception. The quality of medical care is usually checked by monitoring, namely, by continued diagnosis. If one deems diagnosis not as the rigidly determined set of procedures that come to an end before treatment begins, but as the system of information processing concerning patients, their actual and potential complaints, and their general well-being (namely, as a system of running medical commentary and consultation and correction in the widest sense), then diagnosis can be used as the institutional means for the rational planning, control and implementation of standards of health and levels of treatment. But how can such a system itself be controlled?

The best test of a diagnosis is the comparison of the predicted course of an illness and the prognosis which the prediction yields (with the help of the theories it employs) with the observed course. This can best be done if a diagnosed person is left untreated. Such things do happen, and quite regularly, for example, in the case of tests for a new cure of some illness hitherto not given to treatment at all: the research is performed on a group of patients compared to a control group; the efficiency of the new treatment is thus tested, as well as the claim that it has no serious side-effects. This also happens in the case where differential diagnosis is costly and deferment of treatment is, for all that is known, both inexpensive and diagnostic. In the latter case, the development of the illness serves as a test, for example, of the claim that there is no urgency for treatment: the claim is refuted whenever emergency conditions develop. For example, a tumor may be considered slow growing and prove to be fast growing. The sooner such error is detected, the better, of course. This depends on controls, and these are wanting.

Most patients have no control over the availability or otherwise of controls or of their use; even the existing controls are inaccessible to them. A comprehensive computer-assisted medical diagnostic service would raise the availability and accessibility of controls, since there is always the possibility of a class of individuals diagnosed alike on the same empirical grounds, with one or two of them evading treatment for various reasons. This kind of control is insufficient: as it is not always available, though when available, it is excellent. A standard program for it should always be operative, but it will not do to rely on its turning up.

A diagnosis may be a mistake masked by treatment--at times with disastrous results. Obvious examples are cases of the proper etiological treatment for one disease that is merely a symptomatic treatment for another disease which urgently demands an etiological treatment; once the two are mistakenly switched, the mistake may be reinforced and lead to a fatality. An example is the treatment of renal high blood pressure as if it were essential; only the re-opening of the diagnosis totally from scratch will reveal the defective kidney--hopefully before the damage to it is irreversible. A brain tumor, leukemia or some other cancers masquerading as a migraine, at times until it is too late, serve as similar examples.

Can such a common yet avoidable error be spotted by monitors? It is always and essentially a question of speed, time and resources. Hence it is always a question of cost effectiveness, and when properly programmed on the basis of a proper cost-effectiveness analysis, a comprehensive computer-assisted medical diagnostic service that does not replace a human diagnostician might be the best available means. (Computers become increasingly cost-effective monitors since they are made to scan many alternative possible contexts for the monitored case.) Yet, it is far from easy to decide what is the proper cost-benefit analysis and what is the proper program. There is no general program for cost-benefit analysis--except for some set limits to what is permissible and a few important guidelines within the limits of what is permissible.

The comprehensive computer-assisted medical diagnostic service may help preventive medicine approximate an ideal: the regular, frequent, fairly adequate check-ups for the whole population. The frequency of such check-ups, the length and content of the checks, tests, examinations and screenings which are to be included in the list of items of a standard general check-up, these are all matters of cost effectiveness.

The cost benefit of preventive diagnostic services for the public is of concern to the public as such and to the citizen, i.e., to all members and to each member of the public. Moreover, it is very clear that these are extremely hard to compute and are subject to methods of reducing costs, e.g., good publicity given to the program by public information services. Furthermore, computers may be used to test proposals to divide the public into different sub-populations for which different check-up systems may profitably be administered. This is often a powerful method of reducing costs of check-ups at the cost of overlooking items that are rare within these sub-populations, thereby raising the adequacy of the publicly available check-ups within the limits of the budget. Total adequacy is always outside the budget for public health and so some oversights are to be expected. The permissible ones are chosen on two hypotheses: that the omitted items are rare enough within the sub-population where they are omitted, and that on a closer approximation the omissions may be repaired relatively fast and at relatively small costs. The second hypothesis is only true if the system is constructed as pliable and monitored to keep its pliability; computers can be programmed to alert the population regularly for some of the rare diseases omitted from standard check-ups, or to add them to the standard lists on occasion.

This, then, is one important guideline: when a method by whose employment cost benefit is largely reduced with the result of suffering to a few victims, most of whom may be compensated, then surely that method should be employed. What compensation principle exactly can or should be put into operation is disputed among welfare economists, who agree that no compensation principle is ideal.

The concern here is not with the general check-up system, but with its employment to serve as a control over the public health system, over the diagnostic system and even over the monitoring system. Such a service, computerized or otherwise, should be instituted; so far it barely exists. Today, results of general check-ups may be used in consultations with family physicians. They can then be compared and checked against old records; beyond this they can do very little. The benefit of the comprehensive computer-assisted medical diagnostic service is rather obvious here: a few very simple test procedures should be programmed to work regularly to compare every check-up with all current ones and with its owner's predecessor. The latter comparison should be compared to similar comparisons; each result, then, should be compared with the average expected results. These are tremendous tasks--even for computers; they should be severely limited to the search of obvious and relatively common danger signs--and ring a bell to invite further study when necessary.

One of the great obstacles to the achievement of these goals is the current--and, of course, highly reasonable--division of medical services to public health, public services to individuals, and diverse specialized services. Here is an example, and a very important one, of the power of the method of systems analysis, and on two counts. First, a comprehensive system is by far superior to a set of systems covering the same area, since the comprehensive system can be programmed to offer certain desirable interactions between its sub-systems; in the present case, the public health service would offer to individuals a monitor system to check the private or public medical services which they purchase. Second, a combined human-machine comprehensive integrated system is best.

The worst in the current system is not its defects. Defects are understandable, especially such that originate in the absence of an all-embracing integrated medical system. The worst is that a defect at times becomes irreparable due to efforts to avoid it in the first place.

To take a simple example, in medical schools one of the recommendations to future diagnosticians is be complaint-oriented. Although this recommendation is virtuous, taken too concretely it leads to side-effects that must be noticed; experienced physicians are aware of it, yet they prefer not to mention it. One side-effect of the recommendation to be complaint-oriented is the enhancement of diagnosis in the traditional narrow sense of preliminary examination prior to treatment, to the exclusion of monitoring and control, thus serving as an incentive to prescriber to recommend and to therapist to administer unremittingly symptomatic treatment, since the stronger the painkilling (or sedating) drugs used, the more efficient the treatment based on complaint-oriented diagnosis and the less the need for monitoring. The more entrenched and problem-eradicating a system is, the less chance there are for its overall reform or readjustment, even in the face of admitted serious defects. A system which insists on making patients well as long as possible may force the sick to seem well till they collapse; it may, more reasonably, frighten patients to stay away from physicians, as Voltaire and Shaw have already noticed.

The recommendation to be complaint-oriented and its unavoidable and undesirable side-effects justify coupling it with the recommendation to be etiology-oriented. Yet the two are often in irreconcilable conflict. Perhaps, then, the combination of the two recommendations is possible and the conflict may in principle be eliminated. It is the recommendation to be complaint-oriented and lead each diagnosis to its conclusion. In other words, to avoid the conflict between the two recommendations mentioned, one must endorse the third: make each diagnosis as thorough as possible. This recommendation is not always very useful, except in a society sufficiently healthy and rich as to diagnose each of its members thoroughly. The only alternative to the making of every diagnosis thorough is cost-effectiveness consideration, and this alternative makes the three recommendations mentioned at best approximations to the more rational one: at times attend complaints, at times ignore them; at times look for etiology, at times ignore it; diagnosis need not be as thorough as possible, but as thorough as the cost of thoroughness permits; and, above all, always leave reasonable margins of time and of other scarce resources, for monitoring and for overall re-evaluation and reform.

Who should re-evaluate? How is the mode of evaluation ever to be reformed? The expected loss due to a neglected complaint, of a neglected etiology, of a neglected thoroughness, may at times be computed (in accord with decision theory). When computation is too intricate or impossible, then the human part of the computer-assisted system should take over and make decisions as to propriety and preference; the ultimate decision, of course, should rest with the patient, yet individual patients face situations which result from decisions already irrevocably taken without their prior consent or even knowledge, all the rules of informed consent notwithstanding. For example, the low cost effectiveness of an etiology may suggest its omission from the standard textbook, from its statistical list of etiologies, so that patients, unaware of its omission, cannot complain that diagnosis was too skimpy as they might if they were aware of it. For another example, it might be a a common practice rather than the standard textbook that hascaused the termination of an individual differential diagnosis; some patients might protest were they aware of the background situation. These complaints are at times extravagant but at times also justifiable: since the public cost differs as a rule from the private cost, ideally, evaluating each cost should be done twice, from the public viewpoint and

from the private viewpoint. Here is an example. The
calculation of the cost of ignoring abdominal pain alters when
the hypothesis that it is caused by a mild digestive disturbance
is replaced with the hypothesis that it is caused by
diverticulitis or an inflammation of the colon. A diagnostician
may narrow the choice between digestive disturbance and
diverticulitis by excluding the possibility of a severe or
advanced stage of diverticulitis, deciding that it is either a
digestive disturbance that will disappear the next day, or a
diverticulitis that will not be severe or too advanced a day or
two later, and therefore, on cost-effectiveness considerations,
prescribes to the patient another visit the next day, thus using
prognosis as a diagnostic tool. Thoroughness may be
recommended in some cases, and it means, then, an
examination of the hypothesis that the case is neither of a
severe diverticulitis nor of an advanced one. In public-health
considerations questions of thoroughness translate to something
quite different, to questions of priorities on the national agenda
of certain tools and services.

Such practices as the use of prognosis diagnostically are
absent from standard medical textbooks. They are learned in
the hospital during medical clerkship, internship and residency--
though young physicians pay the unnecessary cost of developing
a sense of guilt which reduces the level of performance and
which can be avoided by open discussions with peers.

Consider the case of the course of the illness taking the
unexpected turn; in our example, the diverticulitis is more
advanced than diagnosed or it has advanced overnight more
rapidly than predicted. Presumably on the next day it will be
operated on, especially if the diagnostician is conscientious and
checks in advance that it is indeed possible to perform an
emergency operation on the premises the next day.
Nevertheless, as the disease has developed unexpectedly, the
diagnostician has made a mistake; this is especially so if the
patient is very young, old, or otherwise infirm, and if the
diagnostician ignores the fact that for the categories of these
patients cost effectiveness considerably deviates from the norm.
The mistake is hardly avoidable; its frequency may be reduced
by public discussions, but it is unreasonable to expect
physicians to volunteer the relevant information even though
the errors are justifiable. Had the monitoring been performed

as a matter of routine, the problem would be much more manageable. Such mishaps can be made much less frequent as a mere by-product of the introduction of a comprehensive computer-assisted medical diagnostic service, which will enable its attendants to re-monitor, read, and analyze the most frequent errors in the field. The fear of censure due to admissible error will also be eliminated by the very introduction into the standard computer program the distinction between negligent and permissible error and the institution of a regular public discussions about the limits of permissible error.

The same reason renders the monitor important for the public as potential patients. Monitoring is a major tool for forecasting possible epidemics; in the absence of monitoring, especially when there is a great incentive to conceal information--as was the conspicuous case of the outbreak of AIDS-epidemics--the news may burst into public consciousness. This can cause harmful and avoidable panic, and lead to the raising of funds for fighting the epidemic high on the national agenda, rightly or not, though perhaps more out of a sense of guilt and in the wish to curtail the panic rather than out of rational considerations of cost effectiveness. Such considerations may invite a moral concern, especially when they prescribe the placing of responsibility for gathering information on physicians rather than on patients. Again, the current ambivalence on this matter concerning AIDS is a relevant example. Individualist ethics prescribes patient autonomy and confidentiality but cannot come into conflict with personal responsibility which is always a social matter. As usual, excess in one direction (i.e., siding with autonomy and confidentiality) boomerangs and leads to excess in the opposite direction (i.e., excessive control). This pendulum movement is detrimental to rational planning. Both poles may be avoided when the public is better informed and invited to participate in the decision making process. Two innovations may dampen the pendulum: the publication of statistical analyses of medical errors and the institution of an integrated human-machine system.

So much for the possibility of monitoring and controlling in both the public and the personal case and the interaction between the two. In a given single individual case of diagnostic service to one customer with the aid of the comprehensive computer-assisted medical diagnostic system, monitoring and control can easily be programmed as standing orders in the system, with the aim of using current general information and general public health statistics, in the light of a few rough cost-effectiveness computations, based on simple decision theory considerations, based, in their turn, on general public health information and statistics.

Monitoring and control are not enough, even though in some respects they may easily go far beyond currently available means. Once a proper comprehensive computer system will operate, data and general information may easily accumulate in it, and hopefully enable improvement of its performance, on both the public and the individual level. This would depend on the following two conditions: first, that there will be no pile-up of useless data, and second, that normally data will be repeatedly analyzed. These requirements invite their own control systems. Before discussing controls further, a word of caution: using the general to check the particular is dangerous; it always allows the general to win, since it generates its own (statistical) reinforcement. This is why the programmed computerized part can only be made to ring a bell: it should never decide. Yet, its ringing the same bell often enough may help initiate some very useful and hitherto non-existent controls, in order to raise the level of public health and the quality of public services.

4. CONTROL OF SINGLE DIAGNOSTIC ENCOUNTERS

Public monitoring is primarily for public health, not for personal diagnosis; general check-ups serve public-health purposes no less than personal ones; in both cases the public service has the character of monitoring and uses statistics as a major instrument. Therefore computers are much more useful in public health than in private situations (also they are much less problematic there). The usefulness of a public health system and of the employment of computers in public health may be statistically assessed, and so the whole matter looks unproblematic. Yet the public-health system is useful particularly in problematic cases, when they have to alert individuals and suggest to them that they also check the statistics or the statistical impressions they have, as well as details not given to statistical considerations. Thus, public control is not sufficient for individual cases, but it is necessary all the same.

The most obvious problematic case occurs when a mistaken diagnosis is masked by a mistaken treatment based on it. Here the individual and the public may easily mislead each other. Hence the need to turn to sources of information independent of each other so as to be able to check them against each other; this way one can also use the private and the public case to check each other.

Diagnosis can also be tested otherwise. There are a few traditional ways of testing diagnosis: they were developed in a commonsense manner, with no planning and with no guidelines whatsoever. Consequently, some defects still await the occasion to be ironed out; the occasion may very well be the introduction of a computer-assisted diagnostic service. For example, the common etiquette of the second opinion will alter by the very presence of such a service.

Test by repetition is always a reasonable option; in diagnosis, this is a second opinion. The traditional use of a second opinion is a *terra incognita*. Unavoidably, there are varieties of cases, from the mere going over the first opinion, to the most thorough and expensive new diagnosis performed with no regard for the initial one. Also, there are varieties of reassessment, from the most cursory and barely useful to the complete repetition of the initial diagnosis plus a careful comparison of the two.

A second opinion differs from a consultation: consultation proper is traditionally a part of the diagnostic process (the diagnostic session adjourns in order to restart after the consultation); a second opinion is a repeat performance. What does the pursuit of a second opinion amount to? *The first move* in a repeat performance is simply to survey the procedure of the first to see whether anything is grossly amiss. Usually it is not, but if it is, then usually the second diagnostician is in trouble. A comprehensive computer-assisted medical diagnostic service may prevent this trouble because hopefully it will routinely eliminate gross errors or ring the alarm. *The second move* in a repeat performance is to select crucial junctions in the differential diagnosis and to re-examine them, perhaps by repeating the tests or by refining the tests. Here a comprehensive computer-assisted diagnostic service can

again be of great use because its statistical data may be used
in the second round as to where these crucial junctions are
better located and how likely are errors there. *The third move
in a repeat performance is not as standard and obvious.* The
second diagnostician may proceed to the selection of an added
hypothesis from the list of available diseases and/or etiologies.
It is seldom that the second diagnostician has a larger list of
possible hypotheses than the first (or else the performance
qualifies as a consultation proper). Rather, when a wider set
of possible hypotheses is examined, it is because the very
choice of a second diagnosis amounts to the readiness to extend
the budget allocation for the diagnosis. This may be a matter
of estimate and here the computer may be of help for making
the judgment. This is so even if it is the diagnosing physician
who invites the second opinion: then the budget has been
extended by the first diagnosing physician, whether as an added
customer of the second diagnostician or as an agent of the
patient; the situation here is unclear, as it blurs the
demarcation between consultation and the pursuit of a second
opinion (also, the two may overlap).

Alternatively, *the third move* of the pursuit of a second
opinion may be most radical: the second diagnostician may
propose that the patient enter the process of a general and
thorough check-up, usually in a hospital. This move amounts
to a recommendation that the first opinion be completely
suspended, without offering a specific second opinion but a
possible range for alternative ones. This second opinion in
itself is a new opinion: it concerns the cost effectiveness of the
whole procedure in the way already discussed in the previous
paragraph. The more drastic the move is, the more profitable
it is to request an explanation for it: it may be a mere cover-
up for the first diagnostician, and then the patient is asked to
pay dearly for the cover-up, and it may be a reasonable
extension of the first diagnosis, which is thus deemed the
reasonable process that has led to the prescription of more
diagnosis. It is in the interest of the patient and of the public
at large to find out why the second diagnostician recommends
the extended diagnosis and it is a misconception of the interest
of the profession to suggest that a cover-up is ever justifiable.
If patients could consult the comprehensive computer-assisted
medical diagnostic service before endorsing the second opinion,

they would avoid a lot of grief. Usually, we trust, hospitalization for diagnostic purposes indicates a rare or extreme condition. This (as we have indicated) is particularly so when the second or third opinions are sought from medical research people rather than from expert practitioners. That is to say, in addition to the possibility of using available knowledge there is the possibility of gambling with new ideas and possibly a recommendation to do so, say, in hopeless cases.

Current conditions impose a crisis every time the idea of a second opinion emerges, whether the search for a second opinion is justified or not. If the case is exceptionally severe it is justified, and the recognition of this possibility sends the patient into a crisis. Alternatively, the search for a second opinion may express the patient's distrust and send the diagnostician into a crisis. These are the most obvious options, and in the absence of a detailed discussion they tend to be the only ones. As the idea of a second opinion is always in the air but seldom in the open, the fear of an interpersonal crisis pollutes the atmosphere of all serious diagnostic sessions. **The choice of an open, interactive diagnostic method should be guided by the need to avert the fear of an interpersonal crisis; to this end informed consent to diagnosis has to be legislated first.**

Hospitalization for diagnostic purposes, especially a diagnostic general check-up, is a difficult topic which is usually avoided, and on the excuse that the distinction between clinical diagnostic encounters and hospitalization for diagnostic purposes is sharp--as the latter is very exceptional. Yet there is an urgent need to know, when is hospitalization for the sake of diagnosis or check-up recommendable? It is no use saying, if it does not help it does no harm: it does harm patients, and in many ways, and it does carry its dangers: the danger is that the hospital staff, unlike clinicians, all too often err on the side of excess rather than on the side of the usually permitted cursoriness. It is the patient who pays for that excess in financial terms (even when insured--perhaps particularly when insured), in terms of discomfort and loss of time and even in terms of risks--of iatrogenic diseases due to invasive diagnostic procedures and/or infections, especially those caused by highly virulent strains of bacteria which reside in hospital wards too frequently for comfort.

Usually, however, hospital diagnosis is mercifully preceded by a cursory diagnosis, and the results of that diagnosis constitute a sufficient guideline for the hospital not to spread its net too wide. This is so if the cursory diagnosis prescribes a thorough diagnosis. At times this prescription is made not by the first diagnostician, but by a second one, as a result of an appeal for a second opinion on behalf of the patient. The case of hospitalization for diagnostic purposes recommended by a second diagnostician as a second opinion, when the recommendation is but a cover-up for the suspension of the first opinion, is quite different in that the hospital is not properly presented with a task to test given conjectures. In that case, things may easily get out of hand.

The cause of the trouble is in the confusion about the role of check-ups, which confusion is deeply rooted in the traditional myth of induction and the consequent conception of diagnosis: a complete diagnosis is a complete check-up and the confusion that is expressed in viewing all proper diagnosis as complete is amplified by the confusion that is expressed in viewing all proper check-ups as complete. A complete check-up must involve testing the patient for every known disease, or at least for every known group of diseases, and for a possible defect of each bodily function. This is quite impossible. There are, therefore, traditionally recognized degrees of thoroughness of a general check-up as well as kinds of check-ups suitable for sub-populations. These ideas are seldom articulated or tested. They do change, but too slowly and seldom explicitly, so that some physicians may be uninformed about some changes unless they learn of them by word of mouth, for example when in a discussion of hospitalization for check-up purposes with hospital staff, or from the medical literature. There is much room for improvement of standards for check-ups based both on improved statistical studies and on improved medical knowledge, whether diagnostic or therapeutic. Speed in checking and disseminating new ideas and information is a major *desideratum*, and this is a natural task for the comprehensive computer-assisted medical diagnostic service.

Traditional standard diagnostic tests are at times part of the initial diagnosis, but at times they constitute controls over it. In many hospitals such control procedures are standard. They are hardly ever tested, however, and useless defunct standard test procedures are known to stay in many institutions for generations on end. Computerizing the whole system, with the use of cost-effectiveness considerations, permits the suggestion of new methods for the testing of such test methods. This may lead to the suggestion of better standards. Such procedures may turn the outcome of a piece of monitoring into the incentive for re-diagnosis or even for the improvement of the standards of diagnosis.

The monitoring system has to incorporate some computer services, and to some extent it has. Yet programs to assist diagnosis and programs to assist monitoring have to be coordinated and programmers in these two areas have yet to come together and integrate their work; thus far this has been achieved only by local initiative. The more comprehensive the coordination, the easier it will be--and preferably with an eye on a comprehensive computer-assisted medical diagnostic service as envisaged here.

The more comprehensive a service is, the cheaper it is in the long run, of course. Surprisingly, perhaps, this is so even in the short run--because some comprehensive controls already are traditional and integrating them in an overall comprehensive system would optimize their efficiency with no further ado. Traditional controls of medicine are the institution of second opinion, the public monitoring system and, finally, the standard clinical survey, standard clinical pathological conference (CPC) and similar general control mechanisms in hospitals, in medical organizations of all sorts, and in diverse national bodies concerned with public health.

The idea of standard clinical survey was introduced early in the nineteenth century by Bichat; it was given currency only at the turn of the century, on the authority of Osler and Cabot. This was the birth of the standard clinical pathological practice. Since then the institution of clinical pathological conference (CPC) has become a normal service to the medical community as a means of control over diagnosis: information concerning the pathology of a disease, obtained in *post mortem* autopsies, is regularly used as a tool in the process of checking of diagnoses. It has recently fallen into some neglect, not because its advisability was ever seriously questioned but because of excessive cost-effectiveness. In his *Clinical Judgment* Alvin Feinstein rightly complains about this neglect, since, as

he suggests, there is still much to learn about the practice of
medical diagnosis in such conferences. Here is the place for the
computer to step in. A comprehensive computer-assisted
medical diagnostic service will easily complement the field of
pathology in checking the process of diagnosis; thus monitoring
will turn out to be cheaper, more dynamic, and more
informative. Of course, pathology is where computers are
already used more extensively than in other parts of medicine,
yet as pathological practice is often an unmonitored control
system, it needs all the more a comprehensive system to
integrate into.

          To this end the ethical aspect of the control of
diagnostic practices should be checked, and by ethical
conferences (clinical ethical conference, CEC) and appropriate
committees. Computerization of ethical considerations should
never be attempted; only some of their social implications
might be implemented with the aid of a comprehensive
computer-assisted medical diagnostic service, e.g., the demand
for patient's confidentiality. This demand appears already in
the ancient Hippocratic oath. Moreover, the demand for
individual confidentiality is not specific to medicine. In
financial systems simple means have been adopted, such as the
guarding of records and of computer consoles, the code-
numbered credit card, plus the computer detective service to
catch computer thieves. The implementation of ultimate
confidentiality within the framework of medical diagnosis will be
highly problematic since it will paralyze the monitor part of the
diagnostic system in the clinic as well as the public health
planning domain. Likewise, though a certain part of the
service should be available to every patient via a personal card,
restrictions must exist for the use of other parts of the service,
such as personal diagnostic records and many kinds of available
financial arrangements.

5. LICENSING AND THE COMPUTER

There is one liberal principle of licensing: it is a matter of
expediency to be applied to cases where the requirement to
license a practice is expected to reduce (clearly and rapidly)
the damage caused by practitioners. This principle should
apply to all licensing of all professional activities, whether

medical or not, and including all medical and paramedical ones. Clearly, this is an open matter, since licensing differs in different times and places, and with varying degrees of success. The considerations of licensing are complicated by different irrelevancies. First, guild interests and bureaucratic interests always support stringent requirements for licensing; these may be undemocratic and supportive of the *status quo*. There are social forces against licensing; these may reflect tendencies that enhance the democratic process and those that go the other way. Second, licensing a service is closely linked with other forms of control. In medicine the requirement for prescription of some medications links control aimed at preventing the abuse of these medications to the controls over medical practices. Here two opposite forces operate. A given medication might be on sale, possibly by prescription only, depending on the public interest and on the interest of the producers and the prescribers. These matters invite study-- necessarily in statistical fashion, and the comprehensive computer-assisted diagnostic service might be very helpful here. Also, to the extent that the service may raise the general level of public education, it may render controls less necessary, or less stringent.

Controls over the quality of products and services, as well as over use and abuse, are usually laid at the door of the suppliers, seldom at the doors of the consumers. This is so since placing controls at the suppliers' doors is more economic because they often specialize. This being a technicality, it is irrelevant for the present considerations. Indeed, when different cheap quality control methods are available they are used. For example, prescription: it is neither supply nor demand but a means of control which itself invites monitoring since it is inadequate. As long as there are fewer medical practitioners than patients placing the added controls on the practitioners seems cheaper. Even the control over the prescribers is inefficient; it is then supplemented by tracking consumers of excessive quantities of drugs--usually obtained by multiple prescriptions from unsuspecting physicians, at times even from tacitly complying ones. This would easily be corrected were the computer service in diagnostic records or in sales of prescription drugs efficiently used: were physicians or druggists required by law to report (in complete confidentiality, of course) to a central computer each prescription or sale, it could easily be legislated that, by a push of one extra button, they preclude such deceit and self-abuse. Under this arrangement it will be reasonable to transfer control from patient to practitioner. (This is not to enter the discussion as to whether laws against drug abuse and their implementation are cost effective.) Another example is the maltreatment of minors. Since by law the initial responsibility for minors is in the hands of their guardians, added controls could be implemented to oversee them as well; both controls and added controls may be replaced, usually together.

The specific aspects of licensing of medical practice often stem from the fact that licensing at present still follows the tradition of two major medical practices, internal medicine and surgery, though, of course, their demarcation is not sharp. (This tradition includes two views about psychiatry, the one locating it within internal medicine, or more specifically within neurology, and the other locating it outside medicine altogether.) Moreover, traditionally, physicians were divided into the college-educated upper class of diagnosticians-prescribers and the lower class of tradesmen trained as apprentices, who treated and sold medications. Similarly, surgeons were divided into educated specialists and apprenticed blood-letting barbers. Their monopolies were sufficiently well defined to be guarded by the guilds and by law.

Traditionally, licensing concerns prescription and treatment, and to date less restrictions are put upon diagnosing than on prescription and on treatment. Oddly, hardware operators in hospitals and clinics are often exempt from licensing, even though their task invites more controls than the services of druggists and anesthetists, yet these require licensing. This is far from satisfactory. All traditional licensing, including matters of prescription drugs, should undergo revision, and through placing careful considerations on cost-effectiveness constraints. Cost-effectiveness computations differ, however, according to the varying weight given by physicians to various medical procedures and their outcomes. These depend at times on the custom current within professional associations. Here again responsible conduct is regrettably limited by group interest. These are poorly assessed, since interests vary within a highly fragmented profession, with no overall guiding principles for the whole population and with no proper, high-minded leadership. Moreover, since the fragmented profession is guided by its customs and leading traditions, it inevitably follows various sets of guidelines, not always consistent with each other and often based on the defunct myth of induction, of scientific theory as emerging from facts.

One famous divergence of traditional guidelines may be mentioned here. Traditionally, surgeons are viewed as "aggressive" and internists "conservative". That is to say, presumably surgeons but not internists regard the estimated cost of their operations as insignificant in comparison with their expected outcome. There are, of course, conservative surgeons and aggressive internists, but they are presumed to be the exception. The impact of the difference between these two medical traditions on medical diagnosis is clear in the case where they are at variance, such as the case of regional enteritis or Crohn's disease, which is a chronic inflammation of unknown origin of the gastrointestinal tract. In the course of this inflammatory disease of the intestine the patient might suffer from a mild bowel irritation, but sometimes the effect is far from being mild and in the course of the disease the patient may be found suffering from fistulae or a bowel obstruction. This ailment might be controlled by means given to internal medicine and by surgical means, i.e., conservatively and aggressively. The problem always exists in such cases whether or not to intervene surgically. In this disease neither medication nor surgery promise cure; they may offer mere palliation. The question of surgical intervention in a case of this disease may be offered different answers depending on whether the consultants called upon are surgeons or internists. Both answers are in accord with standards of responsibility of their respective professions. They differ from each other chiefly in their cost-effectiveness estimates. Since these are seldom aired, there is hardly room for critical discussion as to which option is preferable.

The situation is commonplace in medical practice and may concern all life-threatening diseases, like cardiac atherosclerosis; it comes up regularly for discussion in staff meetings in hospitals, in medical journals and other professional forums. Yet to no avail: over-fragmentation and the oblivion to cost-effectiveness considerations prevent critical discussions. Young and old doctors are traditionally considered prone to be respectively over active or "aggressive" and over passive or "conservative"; but at least in departments where they can meet, there is less of a barrier between them than differences between customs of different specializations, including differences in cost-effectiveness assessments. When the same is said of surgeons and practitioners of internal medicine, the problem is all the more difficult to solve.

The problem is general and can be profitably solved with the aid of computers. The most valuable and objective contribution to the theory and practice of implementation that computer services can make is by forming a bridge between specializations by rendering considerations more formal and more amenable to comparison. Turning the present discussion formal may help transcend the labels "conservative" and "aggressive". The ultra-cautious old internist may then turn out to be not less "aggressive" than the impatient young surgeon. This will make the practice of licensing more meaningful as it will impose clearer and more rational standards of care on practitioners. And these medical standards may need to be regulated by an independent public body, aided by a comprehensive computer-assisted program. Though there is a limit to formalization, the computer may serve up to that limit, and even at times beyond it--by prompting humans through the mode of consultation that it may encourage and the monitoring function that it may provide. This kind of service makes the practice of medicine more responsible, and even pioneering in the area of licensing professions.

There is a limit, however, to standard cost-effectiveness considerations and their legitimate use in medicine, and it is often crossed by physicians who are known as miracle workers of all sorts. These regularly use the placebo effect in very large doses `and they thereby may achieve some short-term success, even though it is at the inescapable cost of serious side-effects. They are notorious everywhere since they must have a large, fast-rotating clientele. Many honest professionals seek ways to reduce the damage caused by these miracle workers. It is doubtful that computers may help here, since these physicians do not violate any law: it is quite impossible to outlaw their conduct, and whatever cannot be clearly stated in a law-like general manner is amenable to neither legislation nor to computer presentation. It is clear that they use loopholes in the standard cost-effectiveness considerations. This proves that physicians do regularly consider cost effectiveness; as long as these considerations are effected intuitively, their methods are not given to the reforms needed in order to reduce the damage due to these miracle workers.

Theoretically the placebo effect is intriguing since it is both ubiquitous and traditionally condoned on the view of Hippocrates that treatment without patient optimism is useless. Moreover, it renders even diagnosis curative in part. To an extent diagnosis should be curative, as it should include explanation and advice concerning routine daily affairs; and the dividing line between this and the placebo is very hard to draw. This makes the battle against harmful miracle workers very hard.

Terminal patients offer the easiest and most obvious cases of the placebo effect as it seems that these patients will not live long enough to have to pay for the placebo's side-effects. It is therefore no surprise that harmful miracle workers are particularly pernicious when they manipulate hopes of terminal patients, as these patients may suffer a different sort of damage. Some of them are told all sorts of cover-up stories; they know very well that they are dying, but dare not interrupt the game even though it has been imposed on them and their families. The make-believe medical attitude requires of dying patients self-deception and focusing on fictitious goals, thus pushing them into an intolerable state and compelling them to live their last days and hours anxiously and/or depressively in lies (at times, to culminate in a crisis just hours before dying). In that case the cost that patients and physicians are required to pay for their unwillingness to face a moral problem is alienation. Physicians are alienated from patients and patients from themselves and the entire world. Nevertheless, the dying person has the right to wish to remain alienated from the terminal illness, especially in order to continue to function and to fight off a depression. In such cases, when miracle workers intrude, the inability of exasperated families and physicians to fight vain promise is quite understandable. Yet the only alternative available then is an open discussion of various courses of medical interventions and their honestly estimated cost effectiveness.

In the most extreme case of dying patients, for practical reasons, an open and detailed discussion may not be possible. This, perhaps, is why in our western society individuals have the right to write their will in advance as to how they wish to be treated in such situations. Unfortunately, however, the poor level of medical education of the public does not allow an effective wording of such wills. Hence they leave the legal situation ambiguous: the medical interventions that physicians are licensed to execute properly are quite specific and general statements concerning "dying respectably" do not meet the challenge of modern medicine. The computer may help raise the level of education of the public on this difficult matter and alleviate one of the most difficult moral burdens on all of those who partake in the modern medical situation. The computers may aid display the clinical scene as precisely as needed for the benefit of patients and physicians alike, so as to minimize communication failures and difficult feelings that so often prevail around the negotiations on decisions which involve life-and-death situations.

Nevertheless, every patient has the right to stay ignorant and to live in self-deception. In principle, ignorance can be (and often is) catered for by prior blanket understanding between physician and patient; not so self-deception: it cannot be explicitly agreed upon, as explicitness is the enemy of self-deception. Hence the computer is of no use here. The general use of the computer, as a diagnostic tool, as a tool of self-diagnosis, as a means for spreading medical education and education for control and for autonomy, may help reduce the disposition towards the preference for ignorance and for self-deception--in matters of health and in general.

Chapter 9
THE HUMAN FACTOR

1. INTRODUCTION

When focusing on the objective aspects of diagnosis, especially when taking the computer as the paradigm of explicit and precise diagnostician, then the importance of the subject may easily be forgotten. In the present case the subjects of both the diagnosing physician and the suffering patient are too important to ignore. An objectivist disregard for the subject may evoke the reaction to objectivism in the form of subjectivism. Is this reaction right? If not, can it be checked and replaced by a better reaction to the objectivist excess? How can the excess best be checked? These questions invite a diagnosis of subjectivism and of excessive objectivism; they invite a cost-benefit analysis of the different philosophical strategies involved. We aim to examine the situation in this chapter on different levels: (1) the methodological, (2) the ethical, (3) the nosological, and (4) the political. We take it for granted that neither the object nor the subject can be suppressed with impunity. Suppressing reality seems easy, but it is most expensive in the long run. Suppressing the subject deprives the whole system if its *raison d'être*. The intelligent search for their combination is the search for a healthy lifestyle.

2. SUBJECTIVISM COMPUTERIZED

The concern of this book is to reduce subjective bias as much as possible by looking at the diagnostic and the computer systems, both separately and together. Experts who endorse this goal and propose diagnostic expert-systems software programs as the means of achieving it, often attempt to formalize the systems as much as possible and then effect a precise translation between the informal and the formal system. This raises the problem, can translation be effected with sufficient precision? This problem is pertinent to diagnostics since so much of its practical aspects, as they show up in clinical practice, are considered not science but art.

The computer languages used in current diagnostic expert-systems software programs is, to a large extent, more conceptual than the usual mathematics-based programs used in scientific research of all sorts; they were invented by students of Artificial Intelligence and to a large extent developed by students of computer simulation. (The "shells" of diagnostic expert-systems software programs are more products of Artificial Intelligence research than medical research; medical information is supposed to be poured into the "shells" later on.) Many common concepts and procedures are standardized by programmers, and learning to use them involves learning rules of translation, at times almost without notice. The *locus classicus* of such translations is the average and most elementary text of modern formal logic where students learn, without being told and often without notice, to translate from everyday language into the basic language of logic. W.V. Quine has protested against the tacit assumption that the translations are adequate. Paradoxically, though his thesis that translation is never fully adequate (his radical untranslatability thesis, so-called) is the subject of hundreds of contributions to the scholarly literature, his protest is utterly ignored.

There is no adequate theory of translation or even of the relative adequacy of different translations. It is hard to know if translation, calculation and re-translation distort information or not, though intuition tells that this gives no cause for concern. Nevertheless, attempts have been made to avoid the process of translation altogether, and they go in both directions: raising the level of precision of common diagnostic language and lowering that of computer language.

Attempts to raise the level of precision of common diagnostic language are part-and-parcel of the general cult of precision. Many scientists and technologists value precision and recommend it with no discrimination and no specific reason: they rely on the known, uncontroversial fact that increased precision was essential to certain highly advantageous developments. Precision instruments are advantageous as well, and they are so named because they could not be built without precision. It looks as if precision needs no defense.

Yet precision is also high redundancy: more often than not, gaining precision is sheer loss. High redundancy is not only loss due to investment in irrelevancies. Sufficiently high redundancy may easily clog all access to a system: a book cluttered enough with irrelevancy is so unreadable that its message may easily get lost. Distinguishing the useful precision from the rest is thus significant; historically, rationalist philosophers were recognized by their love of precision; but even the best advocates of precision among them, such as Kant, did not explain. His English disciple William Whewell did and so did later Karl Popper: theories, calculations and instruments should be accurate enough for error elimination.

This answer is satisfactory within pure science, regardless of whether one agrees with Whewell that error-free science is possible or with Popper to the contrary. Moreover, in pure science the required level of precision is easily fixed as that sufficient to test current theories; this is not so in technology. As it is agrees that error-free technology is impossible, the need for a further specification as to the level of error tolerable in technology is also agreed upon. These were never given by any philosopher. Fortunately there are standards for the degrees of permissible responsible error in matters technological; usually they are legally determined and publicly monitored in all advanced democracies. Legal systems are highly problematic, of course, but they are binding anyway--until legal reform improves them.

Whereas scientific tests are means for detecting errors, and precision is but a part of the test procedure, technological testing is the search for side-effects, legally required prior to the granting of permission to implement and/or market; precision is a tool for finding intolerable levels of side-effects. For example, the level of harmful radiation permitted (as sufficiently low to count as negligible) decides the level of precision of tests of video screens of all sorts for their level of harmful radiation. The thoroughness of diagnosis, which may be viewed as its level of precision, depends on errors diagnosticians are permitted to make and the frequencies with which permitted errors are tolerated.

The *locus classicus* for the demand to increase the precision of diagnostic language is a recent volume by Alvin Feinstein, *Clinimetrics*, (1987). Feinstein registers two complaints: first, on the absence of an established system of quality control in matters diagnostic; second, on the loss of diagnostic data that are not reproducible, inaccurate and not worded in standard terminology (*Preface*). Regrettably, however, he neglects his first, important complaint in preference for his second: he proposes a standard terminology to enable computers to preserve irreproducible raw diagnostic data (e.g., the patient looks...), their degrees of strength (e.g., none = 1; light = 2; medium = 3; severe = 4), and degrees of personal conviction regarding them (e.g., 0 = disbelief; 0.5 = doubt; 1 = utter conviction). (A compound report might then be, "in a rather doubtful observation the patient looks to me to a medium degree pale".) The computer's requirement for standardization is easy to satisfy; scientific medical diagnosis requires more: reproducibility. Irreproducible data will not be standardized by standard terminology; the subjective use of standard terms is the production of subjective results, not of diagnostic facts (as the greatest precision and realism in a dream are still dreamlike), since the meaning of a term used for one specific irreproducible event is irretrievable.

Attempts to reduce the level of precision of formal language has a longer history. Aristotelian logic is a formal system. The traditional mediaeval theory of definition required that classifications be sharp--with no vague borderline cases and no overlaps and no residues (this is known in modern parlance as partitions). Yet borderline cases will not vanish. Even the borderline between the two genders is not empty. What of it?

The currently popular, though controversial, theory is Zadeh's fuzzy sets theory, so-called, now advocated by Ove Wigertz in an editorial of a learned periodical for medical information. Now fuzzy sets are sets with probability assignments of set memberships. This, for example, seems to permit describing the borderline gender cases as persons whose gender is a number between 0 and 1 where being male or female is the number 0 or 1. Probability assignments may be variably interpreted. Subjectivists like them to mean, undecided cases: I do not know whether patient x is male or female, so I say patient x is male with probability 0.5. But in reality patient **x** is not half-male. Alternatively, patient **x** may be a normal population, half of whom are males. Alternatively, patient **x** is a rare borderline case. Being a 0.5 male, then, is not a frequency nor a degree of confidence but a fact about patient **x**. But for this borderline case Zadeh's theory is most unsatisfactory on all accounts: borderline cases may be androgynous and they may be with undeveloped genitals; either way, their descriptions are quite categorical, not matters of probabilities. Fuzzy sets, then, are redundant, as it is better to partition a population by gender to males, females and borderline cases, and refine, as the technical term goes, the class of borderline cases into further partitions depending on current knowledge.

Now this suggestion may be unsatisfactory, since we often are in doubt and current knowledge may be insufficient to relieve it. At times current knowledge helps decide whether the proverbial bearded woman is female proper or a borderline case; at other times it leaves this question unanswered. Yet research begins by describing facts as precisely as possible, and if they are vague, this is what their precise description should inform us. A single description of vagueness that fits all cases of vagueness, however, is obviously quite impossible. The systems approach is a commonsense proposal to treat vagueness in a number of ways which have been followed in the past fairly successfully. The systems approach is but a heuristic device: there is no saying how long its success will last.

3. THE COMPUTER AND THE MALINGERER

Every pattern-recognition process must include considerations of noise and at times the source of the noise, since usually the process also includes attempts to consider the possibility that what looks like noise is an overlooked information pattern, and finding the source of the putative noise helps decide such matters. For example, during an examination of brain electric activity what looks like noise may be some electric activity of cardiac origin; only a proper diagnosis will render this possibility an observed fact. Conversely, the view of something as noise is always one additional member of a list of possible hypotheses; it is thus amenable to differential diagnosis, and is often eliminated very early in the diagnostic process, at times to reemerge. The hypothesis in question is that the patient is the source of some interference with the diagnostic process, perhaps while anticipating some result of the diagnosis, perhaps while malingering.

Hippocrates deemed all patients potential malingers. "Keep a watch also on the fault of the patients, which often make them lie about the taking of things prescribed"; the physician should not disclose present or future conditions, he suggested, but try to divert the patient's mind. The physician might be in a better position, it seems, by getting rid of the patient. One might even go further and hope that computers might serve medicine well along the lines drawn in the Hippocratic decorum: only well-formed diagnostic entities, i.e., patterned sets of symptoms and signs, collating themselves nicely into diseases and etiologies, will enter the diagnostic process, with the patient as a mere *locus*.

This looks wrong. No objection to the elimination of the patient need be registered were this elimination a tool helping to eradicate disease; but at times this is impossible. The classic example illustrating this impossibility is malingering: when the patient is eliminated, so is malingering. Malingering as noise stands as a metaphor for ambiguities of all sorts: by definition, malingering is on the borderline between definite pattern and sheer noise.

(From the formal viewpoint, when communication is viewed as pattern recognition, there is no difference between noise and intuition. Between them, these three characteristics-- pattern, noise and intuition--cover all communication. The computer is the metaphor for pattern, the malingerer for noise; intuition has no metaphor since its range is the broadest: from a mother's most common animal comprehension of her offspring to the rarest loftiest spark of genius. From the formal point of view, regardless of anything said or done about intuition, the outcome of all intuition still is noise to begin with and some pattern given to recognition once incorporated in the system (of public knowledge). It should be remembered, however, that even in mathematics the formal point of view is too narrow and it can never be inclusive. Since excessive noise removal in diagnosis is objectionable because it may harm patients, *a fortiori* purging intuition from diagnosis is also objectionable for the same reason. Yet the study of Artificial Intelligence aims at the replacement of the human diagnostician by an inanimate computer, which of course would eliminate intuition altogether.)

The dependable elimination of malingering, if it is not the elimination of all patients, then it certainly is the elimination of all symptoms. After that elimination is achieved, diagnosis must rest on signs alone. In some quarters this is an ideal of scientific objectivity. In those quarters, fully computerized medicine is an ideal too. Admittedly, symptoms are so often aggravating and this may stimulate efforts to replace them, by the construction of artificial tools for eliciting some clear-cut signs. The EEG, and more so the ECG, and blood tests for many types of auto-immune diseases are such tools, though they were invented not just to replace symptoms but to outdo them. Although some (clear-cut) signs exist that replace some (ambiguous) symptom, not every symptom is already replaceable by some known clear sign. The ideal has not been realized as yet. Acting on the ideal would penalize all of those patients who cannot as yet be efficiently diagnosed by signs only; the penalty would be a process of dying accompanied by no known clear-cut signs though quite possibly in great pain. Even when clear signs are easily available, symptoms are not fully dispensable with, since diagnosis usually begins with symptoms and hardly ever with signs: but for symptoms there would seldom be signs other than people fainting in the middle of the street.

To counter these two objections it may be suggested to ignore symptoms only after succeeding to replace them with appropriate signs. This is much better but still unacceptable: some symptoms can be turned partly into clear-cut signs--and fully so only *post mortem*.

This argument has brought new proposals from the followers of the famous behaviorist psychologist B. F. Skinner. They propose that every symptom be taken as a hard fact and dealt with on that level. That some symptoms mislead is for them neither more nor less problematic than that some signs also mislead. They go so far as to propose that the very distinction between signs and symptoms is but an excuse based on paternalism, namely, an excuse permitting the physicians to overrule the declared desires of their patients: physicians consider patients too ignorant of their own best interests and undertake to serve that interest as they understand it, even in the face of objections from the patients.

On Skinner's view malingering is impossible, since he considers all conduct signs, given data. It is not that the Skinnerians endorse all behavior patterns as equally felicitous, but that they find it appalling that anyone except the owner of any given behavior pattern should judge that behavior pattern felicitous or not; their owners, however, may like them or not, and then wish to alter them or not.

Skinner has indeed invented a technique for changing behavior patterns, a technique known as behavior modification. (Only reproducible data are viewed as possibly conforming to patterns and thus alterable.) Not surprisingly, all physicians who share Skinner's penchant for externalizing psychological facts to the point of collapsing symptoms to signs have immediately endorsed his technique. Physicians who deem psychoses purely somatic, recommend to psychotics organic therapy and behavior therapy on top of it, and to non-psychotics they often recommend behavior modification alone. Behavior modification and organic therapy possess a shared background neither in diagnostics nor in therapeutics but they do possess a shared background in philosophy, at least a penchant--a penchant for viewing computers as more human than humans. Or is it the other way around? It is, of course the collapse of the two on each other: *l'homme machine*.

This collapse has excellent cost effectiveness as its practical attraction: at times it is or looks extremely cost effective, and then it is commendable. Examples are the technologies which successfully replace symptoms with signs, such as the machines which check brains for epilepsy. Many simplistic mechanist ideas, such as that the heart is nothing but a pump, are wonderful. Many fantastic mechanist ideas, such as that human brains include memory banks, were wonderfully revolutionary ideas. They all follow the collapse of organism and machine on each other and are highly efficient or cost-effective. The cost-effectiveness, however, is not across the board: the conceptual collapse itself is a technique whose domain of applicability is severely limited, largely by our ignorance. Applying it beyond the narrow domain of its known applicability is at times highly objectionable. Even in research, mechanism is these days better replaced by systemism (as discussed at great length by Mario Bunge).

When there is a symptom or a sign which physicians do not know how to read, the collapse of organism and machine may itself lead to an explanation of the type known as the null hypothesis: assume the unexplained signs and symptoms to be sheer noise. The null hypothesis is always the one to consider and to test first--hence its name: it is the most obvious and commonsense option. When the test result goes against the null hypothesis, however, then it is time to deviate from it. In practice the null hypothesis must give way to some null safety measures: the null hypothesis that a lump in the breast is a mere cyst must give way to the hypothesis that it is malignant, which is the alarm null hypothesis as long as it is impossible (given the current state of knowledge) to name anything else first. Here research and practice diverge: the task of research is to find novelties and it adopts as a null hypothesis the view of a find as common--by describing it as mere noise; in diagnostic practice the task is to ring an alarm and it adopts as a null hypothesis the view of noise as a message of stress--by describing it as a risky pattern. Therefore, the collapse of human and machine is possibly more useful in research than in practice, where it may be clearly dangerous as it diverts our attention from ambiguous data as noise rather than make us try to read them as danger signals.

Noise conceals real patterns and, however rarely, conjures some phantom patterns. Hence, every time a rare pattern occurs, such as symptoms and signs of a rare disease--whether a disease out of its habitat or two or more unrelated simultaneous diseases--they may very well be noise, yet prudence requires to examine such possible noise on the assumption that it is a veritable danger signal. This is particularly relevant to the fully computerized medical diagnosis, which will systematically declare possible noise actual noise, since both rare events and noise are recognized statistically as rare signals; the same computer may be programmed to be systematically suspicious, and this is too costly as well. The alternative advocated here is computer-assisted diagnostic service with humans in control of the readiness to seek danger signals even beyond what a computer is programmed to detect.

A more pertinent possibility of noise lies in the inconsistency of the symptomatology. Computers cannot handle inconsistencies the way we often do in daily life: formally all contradictions are equivalent. Computers are, however, capable of being programmed to discard certain kinds of inconsistency as sheer noise--depending on the context of their programs. Ambiguous items, whether noise-generated or not, are likewise cancelled by computers as a matter of routine, simply because these items add nothing to the statistics they should be relevant to, or because they may suffer contradictory interpretations and then be eliminated as inconsistent. When a symptom is not a clear-cut sign it may be taken as an ambiguous item, and then as a possible inconsistency; it may thus be considered noise and be eliminated. The elimination, reasonable in research, may prove dangerous in practice. A vague tummy ache may be a danger signal for acute appendicitis and physicians should examine patients accordingly.

A mixture of all these cases, of inconsistency, ambiguity, vagueness, conflicting (danger) signals and more, are often seen together--and then read as a sign of malingering. The fully computerized diagnostic system must eliminate all malingering. This would be a relief were the malingerer sheer ballast. Yet the computer may easily eliminate every possible or borderline malingerer as a genuine one, and quite possibly Hippocrates' observation is true that we all malinger a little under the strain of genuine physical suffering, the threat of which is often revealed in the diagnostic session, or the stress of genuine social pressures. Then, only the ideal and the near-ideal patients will escape being regarded malingers; the ideal diagnostic system is left with ideal patients (ones who are precise textbook cases), near-ideal diagnosis is left with near-ideal patients, and vague diagnosis makes us all malingerers to some extent.

This is not to advocate the vague situation as it stands; there is no need to vacillate between the optimum and the current state of affairs. A properly programmed computer available to patients may help them slightly raise the quality and precision of symptoms and greatly reduce the noise.

The computer-assisted diagnostic service here advocated should not be expected to reduce the noise of the malingerer considerably. The noise reduction is due to the education of the cooperative patient; in the malingerer the outcome of education may easily go the other way. A physician will then take proper measures to reduce noise--and the malingerer's complaint--and malingerers will respond by seeking new ways to draw physicians' attention, as is characteristic of their behavior. Physicians may find this extremely frustrating. This frustration is a counter-noise and counter-noise is noise. Transaction analysis may thus come into use here to diagnose the frustrated response as pathological. Frustration is but an etiology; the search for "something else", when characteristic of a patient, is the very case that impedes the study of diagnostics, theoretical or practical. It is the conduct of the diagnostician who seeks "something else" that the malingerer emulates. That search is a deviation, however, when it should be motivated by concern for physical health yet is motivated by something else--by the desire for attention (malingerer) or for inattention (physician), as the case may be. Malingering considered as this kind of deviation, is neurotic or metaphoric. It is rather significant, in either reading, not to take the case lightly, no matter how rare it is. Already Thomas Szasz has argued for its significance, since, he proposes, eliminating malingering--in both physician and patient--will of necessity render its deviation from the social norm (any social norm) and a matter for sociotherapy, not for medical treatment proper, whether psychiatry or somatoiatry. His proposal, he says, will annul psychotherapy altogether and will greatly improve somatotherapy.

We have reached a total impasse. The malingerer will not be eliminated since all patients malinger to some extent and since malingering patients may infect diagnosticians. Will computerized diagnosis help here? We do not think so: if it overlooks malingering it is too bare; otherwise it is too saturated. Either way it is useless.

The way out, perhaps, is the comprehensive computer-assisted medical diagnostic service here recommended. It is recommendable for several reasons, each of which might moderate the difficulty at hand. The service might allow patients to assert and improve autonomy. The information it offers could be publicly available; it could be generally instructive--on matters both general and specific. This may enhance hypochondria and assist malingering; but it may also do the opposite. By increasing ranges of possibilities it might minimize physician's frustration and thus reduce the normal disposition to malinger; but it may also do the opposite. Frustration and polarization between physician and patient may return with a vengeance, and then diagnosticians may be tempted to use specific diagnostic encounters for purposes that are of no concern to the individual patient undergoing diagnosis, whether private or public. Diagnostic deviation from the norm (malingering in some metaphoric sense) is a process which computers both facilitate and hinder.

The moral of the story is that human cooperation with and control of computers is indispensable: so is the readiness for improvement which all the humans involved--physician, patient, the public--should display. Malingering is a deviation-- rare in pure form, common in small measures in all sorts of mixtures; it invites special treatment. Here malingering was introduced as an example of the limitation of any diagnostic system--it is a kind of short-circuit, of course, since the malingerer uses the system in order to undermine it. It is a limitation inherent in every human system, and one both limited and amplified by the computer service. It proves the impossibility of successfully ignoring all symptoms, or replacing all of them with signs. To conclude, all these points are the same. A slight generalization of this same point is, sheer malingering and full cooperation are both ideal, limiting cases. We conclude with our view of both diagnosis and treatment as aspects of one process and with our proposal to initiate a system of comprehensive computer-assisted diagnostic service to implement this view.

Hippocrates' observation of the prevalence of malingering suggests that physicians should anticipate it and stay in control when they meet it: they should develop some personal methods of calming patients so as to minimize it and and they should develop strategies of handling situations in which it gets out of hand nonetheless.

4. DIAGNOSING FUNCTIONAL DISORDERS

Many institutions of scientific medicine owe their character to the classical methodology of extreme empiricism which demands that all medicine should be based on scientific biology, with all physiological facts based on anatomic ones, and so on. After Charcot had failed in his search for anatomic correlates of hysteria, the hope shifted from anatomy to microanatomy, to pathohistology, pathophysiology and pathological biochemistry :

there must be minute brain damages responsible for mental illness: *no psychosis without neurosis*: no mental illness without nervous illness, i.e., brain damage. (Later Freud demarcated psychoses from neuroses, thus changing the meaning of these terms.) This led to studies of types of brain damage, beginning with those of Broca, who discovered areas of the brain where defects cause specific functional disorders, such as the different forms of aphasia. Studies of epilepsy followed, and Jackson illustrated its organic (systemic, not mechanical) character. So did the studies of some specific mental illness clearly caused by some organic fault, such as the dementia in syphilis and Alzheimer's disease. Studies of chemically induced psychoses, from those due to Hashish and cocaine to alcoholic delirium and L.S.D. hallucinations have joined this group of disorders.

The success of these studies, resting on the traditional view of anatomy (or at least physiology) as basic to every psychological phenomenon, was recognized by the initiation of their subject-matter to medicine proper--to be classed as psychiatric disorders, and in some cases (such as epilepsy) even to be classed as neurological. On doubtful authorities such as Kraepelin's, the profession still unequivocally supposes psychosis to reflect mere brain damage. Freud labelled non-psychotic disorders neuroses and recommended that they should be taken out of medicine altogether. Nevertheless, both psychoses and neuroses were classed under the larger category of functional diseases, which is convenient when the borderline between neuroses and psychoses is questioned, or when the borderline between physical and mental illness is questioned--as they both regularly are.

The grave difficulty of approaching mental illness is somewhat reflected in the International Classification of Diseases (*ICD9*, WHO, Geneva 1977). This is the World Health Organization's famous publication that offers a system of classification and standardization of names of diseases for statistical purposes. An exceptionally weak part of this publication is in the area of psychiatry: it uses names for mental illnesses which are too vague. It therefore supplements these names with very brief descriptions of their meanings. What is the nature of this difference between the classification of physical and mental illness? What are these brief descriptions symptoms of?

Unanimity in science is essential for standardization, since science knows no central authority. Yet, contrary to popular belief, science thrives on diversity of opinions and on critical discussion conducted in accord with rules generally endorsed in any democracy. Because the unanimity and the diversity of science are so obviously divergent, most philosophers of science do not know what to do with them-- indeed, none of these philosophers except Popper and those who have learned from him one way or another. The diversity in science concerns theories, as explanations (or possible explanations) of facts, and unanimity in science concerns repeatable facts to be explained. (The requirement for scientific unanimity concerning repeatable facts and the readiness to tolerate theoretical disagreements go back to the foundation of the Royal Society of London in 1660.) ICD, being a matter of standardization should then concern repeatable facts only.

Are there repeatable facts in medicine? Clearly, yes. Are disease entities decidable by repeatable facts? Again, clearly yes. However theory-laden reports on observed facts are the appearances of disease entities; their diagnoses and their etiologies all clearly possess the status of repeatable facts. Similarly, however empirically confirmed our views are of nosology (which is the theory and classification of diseases) and disease, they are hypothetical and thus possibly controversial. Though what really causes malaria is still theoretically highly contested, the matter itself, the factual aspect of the diagnosis and etiology of malaria, the malaria itself, is unanimously accepted as a repeatable fact. In malaria the relapsing fever is recognizably diagnostic, and the specific *plasmodium* organism is recognizably etiological.

Factual information is revisable. For example, malaria has split from one to three strains. Pneumonia has split to bacterial, viral and other kinds; diverse pneumococcal and staphylococcal bacterial pneumonias took the place of the older one strain of pneumococcal pneumonia; and there are now two kinds of herpes, I and II. Also, there is no consumption any more and no neurasthenia any more, and there are new illnesses now, such as acquired genital herpes and acquired immune deficiency syndrome (AIDS). Each of these changes has been publicly acknowledged after the establishment of unanimity concerning new general facts and their repeatable observations.

Is conversion hysteria a general repeatable fact? Yes. Similarly, are paranoia, childhood autism, sleep disorders and ticks (e.g., Tourette's syndrome)? Clearly, yes. Is the status of these as general repeatable facts problematic? No; as general repeatable facts they are not contested, though their boundaries often are. Are there more uncontested general repeatable facts of mental illness? Perhaps, depending on whether epilepsy is viewed as a mental (psychiatric) disease, and nowadays it is not; depending on whether *delirium tremens* is viewed as a mental (psychiatric) disease, and nowadays it is. Is kleptomania then a disease? Since drug abuse and gambling are repeatable facts, are they symptoms? Or are they perhaps diseases? Suppose they are diseases. Is then the Soviet practice justified of locking up dissenters in mental wards?

The problem with the Soviet Union's most deplorable practice may take us to *Diagnostic and Statistical Manual of Mental Disorders-III* (1980) of the American Psychiatric Association. It was intended to serve as a complement to the *The International Classification of Disease 9*. Their naming all the items on their list "disorders" rather than "diseases" will make it impossible to read off their list any decision as to the possible legitimacy of the Soviet practice of locking-up dissidents in mental homes; but a little effort almost takes us there all the same. Disorder Code Number 301.70 of that manual, anti-social personality disorder is described by five criteria, all very complex. Four of these fit many Soviet dissidents. Thus they almost qualify for the status of ones who suffer a distinct psychiatric disorder.

This may reflect an embarrassing, serious difficulty, specific to psychiatry. The kind of problems which can beset *ICD9*, which was intended to standardize names of diseases for statistical purposes, does not include problems related in any way to the Soviet practices. *DSM-III* has a different task, the task to be "first of all, clinically useful, while also providing a basis for research and administrative use" (p. 2). Can that "basis for ... administrative use" prove handy for Soviet authorities? This is possible, as the Soviet authorities conduct towards dissidents is clinically related: the manual makes it easy to view all dissidents as sick. The embarrassment is in the view of deviants as requiring straightening out, not in naming this need by one term or another. This embarrassment is made explicit in the recent revised version of *DSM-III*, known as *DSM-III-R*, American Psychiatric Association Publication, 1987. Among "its basic features", admits the introduction explicitly, is the exclusion from the list of mental disorders "deviant behavior, e.g., political, religious, or sexual", and any "conflicts that are primarily between individual and society". In accord with this, the characterization of anti-social personality disorder is now somewhat altered with only four necessary criteria instead of five, and with the exclusion of Soviet dissidents now becoming somewhat less problematic. Though the exclusion of dissidents from the category of mental patients is now made clearer, regrettably only by some *ad hoc* means, the clear attitude toward them is commendable. The *ad hoc* nature of this attitude, however, shows that something still is amiss. Rather than challenge the American Psychiatric Association, it is more interesting to learn this. (1) What difficulties do its members face? (2) How does the Association hope to assist them with the manual? We will return to the first question, but only after reporting on the second.

Whereas *ICD* lists diseases in all of its editions, *DSM* does not. In particular, *DSM-III* lists (psychiatric) disorders. Is this a significant difference? There are two possible reasons for preferring disorders to diseases. One is that diseases are more abstract than disorders; the other is that a disorder is anything from a symptom to a specific disease with a specific etiology. It may be argued that the field of diagnosis is much more advanced in its physical sub-field than in its mental sub-field, so that more caution is required when classifying mental complaints than physical ones. If this were so, then the difference between disorder and disease would be simply technical, hopefully soon to meet with improvement.

Too much subtlety is concealed here.

Admittedly, terminology is not binding; the use of the term "sociopathy" to designate a disorder is not objectionable, though a disorder is a symptom/syndrome/illness/etiology. But why does *DSM-III* replace "sociopathy" with "anti-social personality disorder"? Why is this done mysteriously, with no discussion of the problem of translation from the traditional terminology to the new? It may sound harsh that as a side-effect of the change in terminology the brave and admirable Soviet dissidents become almost sociopaths even if their certification is still unwarranted.

The replacement of symptom/syndrome/illness/etiology by disorder has an inner logic that should be noted. In a sense, almost everyone agrees, there exist no disease entity ever, only sick people. That is to say, disease entities as such (as *qualia*) do not exist, but as characteristics of sick people they do. Likewise, colors as such also do not exist, nor do personalities as such (as distinct from the qualities of some person), or wisdom and folly as such. Yet in the sense that the blueness of the sky is an uncontested objective fact, malaria being a disease is equally uncontested. That some people suffer from difficulties which prevent them from normally sleeping is also uncontested. Nor is it contested that the etiology of the sleep disturbance is viewed sometimes as physiological and sometimes as psychological. (Stress, depression and asthma also cause sleep disturbance; their etiologies are likewise variably physical and mental.) Unless specified, sleep disorders are non-specific: their status is akin to that of fever rather than malaria. *ICD* lists malaria, not fever; hence, *DSM-III* is not closely parallel to *ICD9* despite its authors' claim.

The practical question here concerns the fate of individuals who display anti-social behavior: among them, who needs therapeutic help and who needs political support? This question concerning mental patients, has a parallel among physical ones: who among the tired is sick? Who among the tired is anemic? It may be hard to draw a clear-cut demarcation, yet it is quite easy to decide that first and foremost dissidents need political help and secondly possibly treatment as well (especially after punitive certification). Why do the authors of *DSM-III-R* show concern over this point?

Because fatigue and anti-social behavior are both symptoms, yet of the two only anti-social behavior is a disorder: despite the progress and increased subtlety of *DSM-III-R* as compared with *DSM-III*, the fact remains: either every case of fatigue, every cough, is a disorder, or else homosexual and anti-social behavior pattern are not necessarily disorders to include in the psychiatric nosology.

We propose that the authors of *DSM-III*, and more so of *DSM-III-R*, are concerned with the search for reliability and the avoidance of controversy. If so, then all their subtlety and confusion could be avoided were they to stick to repeatable observable facts as uncontestable rather than to the vagueness introduced by the term "disorder". As long as decision on the reliability of any claim is to be based on repeatable facts, it is better to rely on them. Is there any other possible way to determine reliability?

This question is open. By definition the facts handled by science are repeatedly observed general facts; hence commonsense facts are different; yet there is such a thing as commonsense reliability, and this is not the place to analyze it. It is agreed that manuals can be useful without making any pretense for scientific validity--the paradigm case is that of cookbooks: they are purchased by cooks because they are useful for them, even though not necessarily for others, and the others need not purchase them. Similarly, *DSM-III-R* does claim that it is (not an introduction but) a useful manual for qualified practitioners. One may ask, then, are cookbooks sufficiently reliable for their recipes to be deemed repeatable? Do some cooks contest recipes? Do some psychiatrists contest classifications of disorders? Certainly. Is this reflected in *DSM-III*? No. The question has thus become, do psychiatric practitioners find *DSM-III* useful in their efforts to diagnose and treat mental patients?

This is a controversial question: it is seriously and significantly controverted that the manual as such helps practitioners, qualified or not, to diagnose mental illnesses or to help mental patients. For, it has all too restrictive an effect on the diagnosing and treating psychiatrist. The following are the philosophical sources of these restrictions.

The sources of the controversy over *DSM-III* are inductivism and mechanism. There are many controversies within the philosophy of science, and some of them concern classical inductivism, the view which disapproves of the employment of hypotheses, or at least it grudgingly approves of very tame hypotheses proposed after the collection of many facts. Yet this view is definitely on the defense and hypothetico-deductivism of all sorts is advocated, especially by those who insist that only reproducible general facts have a place in science (since this prevents them from being free of all hypothesis: reproducibility is always the theoretical element in every observation report, since only single reproductions can be observed). To repeat, the demand for the reproducibility of facts is the only universally recognized requirement within the whole scientific community. It is now clear that *DSM-III* is an exception to this, and may safely be declared pseudo-scientific in the following distinct sense: *DSM-III* rests mainly on commonsense reliability in lieu of scientific, repeatable, general facts. Some psychiatric facts are repeatable--such as childhood autism and Tourette's syndrome. Some are mixtures of scientific and commonsense facts. This may be utterly unobjectionable--at least as a starting point--yet the hard fact is that *DSM* has three or four versions, very few of the variations being rooted in either new fact or theory. *DSM-III* is different from its two predecessors chiefly in its presentation of matters in as standardized and formal a manner as possible, as well as in its utter suppression of all mentalist theories which are at the root of psychiatric classification, diagnosis and treatment. Thus, *DSM-III* avoids controversy in matters theoretical at the cost of losing much diagnostic and treatment information without discussion. This loss is supposed to be balanced against the gain due to the standardization that the revised version of *DSM-III* offers. Where is the gain?

Standardizations of mental disorders are often matters of tradition. Alzheimer's disease and alcohol intoxication are clear physical syndromes yet they are considered psychiatric, not because they have mental symptoms as well, since almost every neurological illness has them, and since almost all illnesses normally produce in normal minds anxiety, depression, and so on. They are considered psychiatric disorders because of some socio-historical factors, factors which have very little to do with the classification of diseases and which anyway will not command scientific unanimity today. At most they will be accepted as conventions for merely administrative purposes.

An outstanding instance of the irrelevance of socio-historical factors to the diagnosis and etiology that command scientific unanimity is the case of homosexual behavior, considered a mental illness since Freud and remained so until pressure from the gay liberation movement forced the American Psychiatric Association to decide, in 1973, to remove it from

the *DSM*. Not quite, though: *DSM-III*, includes, under Code
Number 302.00, Ego-dystonic Homosexuality; the absence of
ego-dystonic heterosexuality there, or even of ego-dystonia as
such, raises the suspicion that the bias against homosexuals is
now less conspicuous but not less present. The absence of ego-
dystonia may be understood as due to the refusal to admit it
on the ground that it is a mental entity. This excuse will not
explain the absence of ego-dystonic heterosexuality, however, as
it must have on all counts the very same status as ego-dystonic
homosexuality, ego-dystonic fetishism and any other ego-
dystonic quality; why then is ego-dystonic homosexuality
singled out? Possibly other code numbers (e.g., 302.70-302.71)
of *DSM-III* do deal with behavior patterns characteristic of ego-
dystonic heterosexual disorders. But if so, why not use
explicitly the two terms in parallel? If the manual's authors
are of the opinion that there is no parallel here, why not say
so? Whatever is the case, it is clearly a case of evasion.

There is a possible reason for this evasion: perhaps with
one exception all ego-dystonias are as yet not sufficiently a
specific observable disorders. Assuming that this is so, then
the authors of *DSM* are in worse trouble than they realize: an
objection should be voiced against the very use of the term
"ego-dystonia" as too abstract and so open to controversy. A
thought experiment should make this clear to anyone ready to
invent disorders as lightly as the authors of *DSM-III*. Consider
"homosexual disorder", "homosexual personality disorder" and
"homosexual ego-dystonic personality disorder": Which of these
three is more abstract and which of them is more
controversial? The answer may lie in the comparison of the
homosexual and the heterosexual cases; we have now to
compare three triplets of concepts: "homosexual disorder",
"heterosexual disorder" and "disorder"; "homosexual
personality disorder", "heterosexual personality disorder", and
"personality disorder"; and finally "homosexual ego-dystonic
personality disorder", "heterosexual ego-dystonic personality
disorder", and "ego-dystonic personality disorder". Without
knowing the exact meaning of the terms one can already
conclude that unless there is a parallel between the homosexual
case and the heterosexual case, there is a bias here--justifiable
or unjustifiable but controversial. This critique should force the
authors of *DSM-IV* to reconsider the situation. We propose

that there is no way out of the predicament but to admit that
only observations repeatedly reported, and as repeatable, are
beyond controversy.

Perhaps *DSM-III-R* classifies Adult Anti-social Behavior
as a "condition" wishing to avoid classifying deviant conduct as
a psychiatric disorder. This, of course, is laudable, yet to be
successful it should show that a condition is not a disorder.
How is this to be done? After all, the mere terminological
change does not prevent a controversy, especially when
unexplained. The reason the term "disorder" was coined, we
have suggested, is that its inclusiveness was expected to
prevent the controversy as to whether it is a symptom, a
syndrome, a disease or an etiology: it is a disorder; but now
it can be a disorder or a condition, and so controversy rears its
ugly head again. For example, we face the question, Why not
consider homosexuality a "condition"? How is homosexuality
considered in *DSM-III-R*? As a disorder when classified as a
homosexual ego-dystonia (and so it is classified together with
heterosexual ego-dystonia under Code Number 302.90, Sexual
Disorder Not Otherwise Specified), but as a condition when
classified together with heterosexual adult anti-social behavior.
(Reference to ego-dystonia is made only in the *Index*, while in
the body of the manual the term is replaced by "persistent and
marked distress", p. 296.)

What is the medical status of the anti-social behavior
that is not a disorder proper yet invites psychiatric attention
all the same? Why should it be the object of psychiatric
diagnosis and/or treatment? According to the authors of *DSM-
III-R*, it may serve as an item contributing to the diagnosis of
disorders proper; why then should it deserve a separate
diagnostic code number? It is not enough to classify anti-social
behavior separately as a problem which deserves psychiatric
attention. The authors of *DSM-III-R* promise to the
diagnostician a diagnostic system based on scientifically
repeatable facts, not a list of commonsense problems.
Otherwise suicidal behavior, too, would have been granted a
diagnostic code number. The authors of *DSM-III-R* may have
observed (p. 359), however, that some people who display
deviant behavior do so with some remarkable consistency, for
instance "professional thieves, racketeers ... dealers in illegal
psychoactive substances". They name this behavior anti-social

and classify it separately from the psychiatric disorders. It looks, rightly or not, as if they perform some slight-of-hand here: the prefaces of *DSM-III* and *DSM-III-R* promise to present uninterpreted facts which are uncontroversial: presumably to that end they have replaced disease with disorder. Yet allowing personality disorders, ego-dystonic disorders and adult anti-social behavior conditions, they render "disorder" and "condition" synonyms for "disease" or else their terminology is too vague to be useful.

5. SCIENTIFIC PSYCHIATRY

On some central issues *DSM-III* and *DSM-III-R* are vague to the point of being misleading--clinically and scientifically. The popularity of the admittedly imperfect *DSM-III* and of *DSM-III-R* is a cause for concern. Norbert Wiener has noticed that social scientists emulate the outer characteristics of natural scientists. Likewise, psychiatrists try to emulate the somatoiatrists to the point of standardizing terminology and offering classifications which makes their techniques amenable to dubious statistics and computer-assisted diagnoses. In this process they deprive themselves of their own psycho-social science-cum-technology.

There is no scientific unanimity as yet ruling a sufficiently large domain of diagnostic facts or other nosological facts in psychiatry to warrant standardization of its terminology, let alone formalization, namely, the introduction of the computer for statistical and/or diagnostic purposes. Yet this idea is clearly implied by the authors of *DSM-III* and *DSM-III-R* in the introductory material and in *Appendix A, Decision Trees for Differential Diagnosis.* As the philosopher of mathematics Imre Lakatos said, premature formalization is defensive, aiming at the avoidance of all problems at the cost of ignoring all challenge and losing the fun and excitement of the exercise. Since there is little difficulty diagnosing dysfunctions, and since disorders are described as observed dysfunctions, there is little left for diagnosticians to do unless they have some etiologies with which to work, certainly nothing to formalize. Most psychiatric etiologies being controversial, they are excluded from both *DSM-III* and *DSM-III-R.* Why then take the trouble? What difficulties have forced the

American Psychiatric Association to invite a revision of *DSM-III*? What problem stood behind the revision?

We have two points to offer, concerning the import of *DSM-III-R* for matters of mere style yet of practical significance nonetheless. The first has to do with the style of therapy, the other with the patient's lifestyle. Concerning both matters *DSM-III-R* seems to us to be partly commendable and partly not. Therapists must be as open to their patients as possible (we take it as self-evident), both in their articulation of problems, doubts, and concerns, and in their determination as to what they recommend (whether their recommendations are accepted or not). *DSM-III-R* helps articulate problems in various human dimensions: biological, psychological and social-- on the condition that its users are consciously problem-oriented.

So much for the commendable aspect of *DSM-III-R*. To discuss its less commendable aspect we should refer to an aspect of medicine hitherto ignored in this book. In addition to symptoms, diagnosis, etiology and prognosis, there is such an item as pathology. To repeat our examples, of syphilis and malaria, in addition to their diagnosis, etiology, etc., their pathologies are brain necrosis in syphilis and anemia in malaria. When etiology is absent pathology looms large. In psycho-pathology, especially in the revised version of *DSM-III*, there is scarcely an etiology. Nevertheless, this version neglects all pathology. We find this regrettable.

Possibly the authors of *DSM-III-R* wished to avoid difficulties concerning the factual status of psycho-pathology and/or the moral status of the psychiatric diagnoses of lifestyles. This may explain their division of psychiatric disorders between what they designate as Axis I, of major psychiatric disorders, and Axis II, of personality and developmental disorders: these concern lifestyles. The division, however, is clinically and scientifically rather limited. Recent difficulties in identifying genetic markers for major psychiatric disorders, for example, indicate that deviant lifestyles are at times expressions of major mental disorders--the way fatigue is at times a symptom of sleep disorder and at times a symptom of anemia. Indeed, there is no clear-cut demarcation between sick and healthy lifestyles. It has been the tradition of medicine to recommend "soft" changes in lifestyle where such changes are medically significant. Their prescriptions, being least aggressive, may serve as an ideal for medical practice in general. The case before the authors of *DSM-III* is different, however. It calls for a differential diagnosis of lifestyles and therefore, says Szasz, it goes beyond the traditional mandate to the medical profession.

We disagree: the mandate to the medical profession is a socially determined and publicly controllable matter. The question before responsible physicians, however, is less concerned with whether a problem is within their mandate than with whether it is within their competence: do they have the knowledge to diagnose and treat it? The intent of many members of the anti-psychiatry movement was not to stop treatment but to change its status, to have it declared non-medical. Though in this they failed, their impact was tremendous in the broadening of treatment to obviously non-medical areas, where ideology and way of life play more important roles than anything even remotely medical.

An obvious example for this change is the official Canadian attitude to mental health as expressed in two Canadian Federal Government Discussion Papers, *Achieving Health for All: A Framework for Health Promotion* of 1986 and *Mental Health for Canadians: Striking a Balance* of 1988, where the balance in the title is both between enjoyment and development and between individual and social interests. We are not able to offer a reliable assessment of the impact of these papers, but they clearly express an attitude merging psychiatric and anti-psychiatric ones. (For more details one may also consult Canadian Provincial Reports.)

Though this trend is conspicuous in psychiatry, it is characteristic of affluent society in general, and in medicine it is seen when medicine embraces nutrition though nutrition strongly relates to lifestyles, from culinary matters to religious dietary laws. We shall return to this in our *Conclusion.* Here, suffice it to say that in the case of problems that pertain to lifestyles, physicians have to recognize the impasse caused and sustained by the prevalent medical style that disregards the larger framework of medical diagnosis proper, a disregard justified on the pretext that the larger framework is by definition external to biological medicine proper. Hence, this medical style suppresses the need for enlarging the clinical context so as to embrace the study of individuals--their mental and biological world--at large, of society at large, of social and political life at large and of lifestyles in general and of the place of illness in them. The limitation of *DSM-III*, as well as of Szasz's critique of psychiatry as pseudo-science, is nothing but the limitation of the prevalent inductivist medical style. At stake is an attempt at capturing this style by the use of computers in psychiatric diagnosis without changing this style. All this is necessarily limited and limiting.

The claim of medicine to scientific character is a great hindrance, especially when coupled with the disclaimer that medicine is not a science but an art. Clearly, not all repeatable factual information is scientific, and not all of medicine is scientific. After all, scientific medicine is one century old and is limited to the modern world, whereas medicine is age old and universal. Moreover, what counts as medicine is a matter of social convention--at time administrative--rather than of scientific fact, even when we confine medicine to science and to scientific technology. This is not to deny medicine, as conventionally understood, its honorific place in society, nor to deny the commonsense demarcation of medical technology as largely patient-oriented.

As far as psychiatric diagnosis is considered, two points are clear. On the one hand, ignoring any unrepeatable fact is itself neither scientific nor uncontestable. On the other hand, computers are useful for storage management and manipulation and arrangement of repeatable factual information. Since medical technology is largely patient-oriented, psychiatric diagnosis should attend to patients' immense need for psychotherapy and the overwhelming and quite repeatable evidence that psychotherapy (with and without biotherapy) is beneficial one way or another. There is not enough research into the matter, as most extant data are not repeated, perhaps not repeatable. Therefore, a great deal of the statistics which circulates in `the literature is invalid, not to say pseudo-scientific. But there are repeated cases of successful treatments which would be irresponsible to neglect. On the whole, as long as responsibility requires treatment it also requires diagnosis and research, so as to render it repeatable. The sifting of extant scientific diagnostic tools from the rest is an urgent task. Psychodiagnostic and psychotherapeutic facts are elusive-- as are psychobiological facts which are and should be closely related to them; the high price of not differentiating science from common sense is severe constraint on the ability to plan and assess treatment rationally. This holds for almost all psychiatric diagnosis and treatment and for some somatoiatric ones as well.

It is this commonsense characterization of medicine as patient-oriented that leads to the suggestion that a convention be endorsed to include the treatment of physiological dysfunctions in scientific medicine--on the proviso that treatment repeatedly prove useful and on the proviso that further research into this matter be undertaken. The same holds for the proposal that all treatments of physiological dysfunctions--be these treatments deemed scientific or not--stay within medicine (which is still to a large extent an art, not a science). It is important to keep separate the time-honored convention of endorsing claims for scientific status on the basis of repeatability and the invalid claims for scientific status on the basis of the claim that certain data constitute raw observed facts. The claim for scientific status made in *DSM-III* and in *DSM-III-R* is based of the latter kind; it is thus absurd. Fortunately, this matters little, since many individuals and groups claim some sort of unearned social status. What is regrettable is the confusion *DSM-III* and *DSM-III-R* perpetuate: confusion between the reliability of all scientific technology, including scientific medicine (which rests on tests according to socially accepted and improvable norms), on the one hand, and commonsense reliability on the other. Since all medicine incorporates both kinds of reliability, both are necessary. Yet keeping them distinct is an important aspect of the scientific tradition which should be preserved--so as to prevent science from becoming so dependent on extant commonsense that researchers will be prevented from making bold conjectures that depart from extant commonsense. For the freedom of bold imagination is the heart and soul of the scientific endeavor. Moreover, the standards of responsible technology are distinctly higher in scientific technology than in commonsense technology. **The desire to raise the standards of medical services is inducement for maintaining the distinction between the reliability of tested and corroborated reports of repeatable facts and any other kind of reliability used in medicine.**

Chapter 10

CONCLUSION:

THE FUTURE OF THE DIAGNOSTIC SERVICE

1. INTRODUCTION

This final and concluding chapter constitutes a somewhat detailed picture of the comprehensive computer-assisted medical diagnostic service as here envisaged, embedded within a broad framework of diagnostics, a crude picture of the partial services as they exist, and an outline of a proposal concerning the possible way to bridge the gap between what exists and what is desirable. Since the project advocated here may seem utopian, it is perhaps permissible to repeat that the accent here is placed on

(1) the computer as representing formal rules,
(2) the practical use of computers as a model for the rational planning of new and reformed medical services, and
(3) guiding principles which may lead to a desired future equilibrium point.

With some effort increasingly many existing services could be computerized to some extent, expand, and merge with other computerized services.

2. THE DIAGNOSTIC SERVICE ITEM BY ITEM

The service here envisaged should be comprehensive in a few senses. It should handle all possible diagnostic fields, combine many services, be linked to other services such as treatment, and be centrally coordinated though not centralized. A free information exchange should remain the chief asset of such a system, not economies of scale. A computer service cannot offer more than the information stored in its banks, the computations that it is programmed to perform, and a combination of both as computations involving and expanding stored information. Presented here is a list of items which can be stored in computer data banks, what can be done with these items and, particularly, what the limitations are on such information according to basic information theory.

The computer-service can offer not only clinical or diagnostic items and computations, but also information about contexts--kinds of information hitherto hardly ever considered. For instance, information concerning standards, diagnostic procedures and their costs and benefits. The service can present, with the aid of the mathematical theory of queuing and storage, some details about medical procedures which are semi-administrative concerning the advisability of hospitalization or of congestions in hospitals, and so on. The computer may be a resource base for information services at large, to physician and patient alike, akin to those available in the computer services which colleges offer to their students and which operate now in large hotel complexes. Here is a socioeconomic bottleneck in many medical services that has become fairly well known as a serious and urgent problem throughout the entire industrialized world. Finally, the service may offer all sorts of simulation which the market is full of these days, especially in training programs.

As to computations, the service can offer statistical computations and cost-effectiveness computations on top of solutions to such problems as diagnosis which may concern, e.g., chemical analyses of body fluids. Statistical computation problems may involve the factors entering the considerations of advisability of a test or of a method of treatment. The cost-effectiveness computations may be differential, i.e., including a long list of different possible causes of a disease as well as lists of given different possible methods of treatment--each tailored to given categories of patients. At each stage of the computation possibilities of development along given diagnostic or therapeutic courses can be computed. All this sounds frightfully complicated, and unquestionably it is for pre-computer situations. This is what makes the introduction of the service so advisable. Any required computation may be performed by all sorts of sophisticated computer programs, such

as those readily available in any university hospital. At
present the diagnostic use of a university computer service is
almost unthinkable. Here both cost-effectiveness considerations
and the need for specific medical diagnostic services is what
should make the comprehensive computer-assisted medical
diagnostic service as here envisaged much more useful and
attractive than any computer service now available.

Computers also increase the fields of choice available to
patients. They should not prevent the choice of an option
usually considered inadvisable, which option may be preferred
for some personal, moral or ethnic reasons known to the
patient alone. The limits of the rationality of the patient's
choice invite considerations and computations specific to that
choice. This is in accord with the current theory of rationality
which gives way to pluralism, yet contrary to current practices
which still ascribe rationality only to domains where scientific
unanimity reigns supreme. (This domain is the narrow field of
scientific repeatable facts alone; it is all too narrow.) Without
going deeply into theory it can be easily and intuitively seen
that pluralism provides ample room for improving each choice
within some framework which is taken as its context.

Pluralism has its cost or cost benefit. Its cost is the
fragmentation of the decision system into many sub-systems.
The benefit is in a certain crystallization or integration--the
package decision coming with the knowledge that a certain
consumer (usually the patient) belongs to a given set, thus
joining the general with the particular not on the highest level
of generality but within a limited context. (This is a standing
technique in systems analysis.) The cost-benefit of this
pluralist situation can be reduced by the computer to some
extent, if and when its statistics should take account of extant
pluralism.

Putting computations in general aside, we turn to
information. The cost of having large statistical and other
general items became so low with the advent of the computer
that there is no need to go into a list of all possible useful
kinds of information. Considerations of cost effectiveness do
remain, however, and they ought to guide computer analysts
and planners: banks must be stocked selectively, with an eye
on cost effectiveness and, when the banks operate to capacity,
also on opportunity cost. The limited capacity of memory

banks and the increasing cost of access to any single item from
an increasingly large bank offer incentives to streamline and to
dump items not frequently used and not very useful. Each use
of any item has a value and a frequency, and their product
should be above a given minimum for it to stay in. This
should be and usually is repeatedly checked. The costs of this
checking and/or of renovating are usually computed. When the
cost is too high, yet the overhaul is still needed, the whole
bank may be better scrapped. This often is the case, and
trivial as it is, it is all too often overlooked.

The idea that some signs and symptoms must be
ignored, so natural to information theory, is traditional
medicine: already in the 18th century the distinction was
made between essential and accidental symptoms--the
terminology is Aristotelian *par excellence*, yet the profession
could not eliminate the idea of accidental symptoms, i.e., of
noise. There is no need to excuse the elimination of allegedly
accidental symptoms as noise; yet, there is always legitimate
apprehension about these matters: What is known that
permits the cavalier dismissal of given symptoms in a given
concrete case? When storage of information is so cheap the
disposition is to store just about every item, even (seemingly)
accidental details (which may turn out to be useful). This is
clearly a mistake exposed by queuing, storage, and retrieval-cost
calculations; usually computer technologists use some rough
estimates and dump data automatically.

Some items, of course, regularly drop out at once. They
are items judged as noise. Yet this judgment is hypothetical:
when noise is very clearly eliminated, it ceases to be viewed as
noise, and is then viewed as some accretion. Even if noise
replaces a signal--e.g., a Morse-type dash is turned into an up-
and-down whistle--it is not really noise but the signal in it is
transformed but still a recognizable variant. To count as noise,
a signal must be readable in different ways. By definition,
noise is an element of indeterminacy, of the lack of clarity, a
misprint not easily to be corrected unless the original is already
known. What can be done with noise? One can guess the
proper original signal it distorts, and one can do so in a precise
or vague manner, where the degree of precision will also be the
degree of possible deviation from the right message. That
degree depends on the surrounding messages, i.e., on the

patterns available to be recognized and their general likelihood, namely on the degree of redundancy of the message. Hence, signals are used to identify or diagnose patterns, but also, when in doubt--i.e., always--the opposite is done as well. The more detailed the message and the more detailed and redundancy-filled the surrounding patterns, the easier it is to correct a distortion due to noise. Redundancy is added to a message in order to make it easier to correct. Shannon's theorem says, eliminating all possible error from a message requires including endless redundancy in it. Given a certain transmission of a message, redundancy may be achieved by repetition. A message containing no redundancy may be distorted beyond recognition by a slight error.

Shannon's theorem is the key to all pattern recognition, whether computerized or not. Its application to formal systems, whether decoding a Morse-code message or performing any computer program, is less problematic than its application to normal situations (whether of recognition or of checking a case of recognition). The theorem holds in the normal diagnostic situation: when in doubt, a diagnostician can add redundancy--a repetition in the form of a second opinion proper, or a second opinion in the form of test, or second test. Tests are by far more effective than the traditional anamnesis, and they have the same function.

Pattern recognition can be divided into four stages: the two first stages are the patterns and the elements; they are the more stable and less often given to drastic revision; they are revisable with the growth of knowledge or the reallocation of resources. The two later stages are fitting and decision; these are problematic on a day-to-day basis and should be kept open to frequent alteration.

This bears out the claim, "medicine is an art, not a science". More precisely, art enters when the limitation of science invites it. It is advisable to keep the art, and have the computer serve the science with ease and thereby help the artistic contribution enter as soon and as forcefully as possible, and exhibit the required flexibility and dexterity in applying the rules of thumb that are not as well anchored in science as one wishes. The growth of science always eliminates old art: old rules of thumb give way to scientific procedure, these calling for new art give way to supplement them. Art need not fear science: when science replaces old art it also opens new avenues for more art.

It is easy to apply this consideration to simple extreme cases, such as a diagnosis leading to a clear invitation for an expensive operation. The cost effectiveness of a possibly redundant operation is the key: four sets of possible costs and benefits should be calculated: the cost and the benefit of performing or of not performing the operation, given that the diagnosis is correct and that it is incorrect. The computer can help reduce the cost of the considerations and reduce the margin of error by increasing precision.

The case of the odd occurrence or the case that is exceptional--according to the accepted theory or according to common sense--is quite similar. Though methodologically such a case is a very important ingredient in every learning process, it nonetheless is filtered out as noise by the clinically defined processes of pattern-recognition programs. Hence, the identification of highly exceptional cases cannot be computed and this is why diagnosis should always remain only computer-assisted, not fully computerized: some standards of fit between general theoretical patterns and specific given cases should be implemented in the programs of the comprehensive computer-assisted medical diagnostic service. This way, when discrepancy arises it will ring a bell and invite an attempt at a revision of a theory, the updating of data, or the changing of the standards of diagnosis or of control. What is usually ignored is--by definition--mere noise. This is highly counter-intuitive. Noise is a signal which detracts from the value of a message to which it is added. Yet noise need not be an added signal: it can be a distortion of one; in the present case the noise is information proper, and even highly valuable and advanced information--indeed, too advanced and therefore not comprehended!

The reason for this counter-intuitive result is the breakdown of a metaphor. The use of the word "information" in the label of information theory here is only metaphorical since that theory handles signs, not symbols--not information proper but its mere vehicles. The usual case is of a set of signs common to transmitter and receiver transmitted over a noisy channel. The channel in the present case is as near ideal as practice requires: noise removal in computer systems (by redundancy) is excellent. Rather, there is no transmitter other than Mother Nature, and She is not forced to use our signs. We impose our theories on Her, and the cases in which She refuses to obey we call the odd cases or the anomalies--they are anomalies only relative to our theories. Hence, we are at liberty to treat anomalies as noise relative to older theories, as messages relative to the new and better theories and as pregnant information in the transition from the old to the new theories. The excessive use of and reliance on the computer will impose the choice of disregard for the anomalous information as sheer noise; it will thus foster stagnation.

Omitting important information as sheer noise is not limited to taking anomaly as noise on the ground that (by definition) it does not fit current theory. A worse situation occurs in a rare case such as the rare disease, which does not fit current practice. The rare case is usually indicated by a rare sign or symptom which will likewise tend to be misread unless repeatedly presented (by second opinion, to stretch our metaphor). Still worse than the case of a rare sign or symptom is the rare combination of common signs or symptoms, which happens regularly in all cases of dual and multiple diseases of independent origins. Formally speaking, a multiple illness is one different from its components. Medically speaking, quite often a dual disease requires more than joint treatment. (There is also the case which need not be discussed here, of a dual disease with a clear etiological connection; the extreme example for this is AIDS, of course, as patients suffering from this disease may become victims of peculiar infectious diseases as well as of cancers. There are many similar examples.)

What can be done about the rare disease and the multiple diseases of independent origins? How can the risk of ignoring some items as mere noise be minimized? Risk can be minimized only when the failure of one hypothesis to fit an extraordinary risky item into an ordinary pattern leads to further attempts to explain it. The failure is not often due to the persistence of the alleged noise: information theory is reasonably new and is still fairly unknown; it is therefore too seldom noticed that the systematic repetition of a sign taken to be noise raises the likelihood that the sign is a part of the pattern carrying a message. Not considering the possibility that the alleged noise is a message, physicians tend to treat it and thus suppress it. When the suppression fails, when the medication does not cure, and if the patient is still not dead and if rediagnosis is attempted, then perhaps by shifting from the patient's one illness to the other, then and only then may the hypothesis surface that suffering is due to two illnesses. With the help of information theory it may be possible to propose a comprehensive computer-assisted diagnostic service that could both offer the hypothesis of a rare or a dual disease of independent origins--as a rare occurrence, of course--and reduce costs of diagnosis to such an extent that the keen and alert diagnostician may, at least, entertain, however tentatively, some alternatives, however rare.

Any rare pattern--a rare disease or a dual disease of independent origins--usually may be safely ignored on the basis of its rarity alone. Ignoring it, however, will make it look ever so much rarer than it is. Cost-effectiveness computations of this increased rarity will tempt the programmer to omit its very mention from the initial standard program. The way to avoid this error is that of careful checking--e.g., *post mortem* operations--and monitoring--e.g., the quick check for a discovery of any discrepancy. Such considerations, however, are part of the fourth stage in pattern-recognition procedures, following the study of patterns, their elements, and the fitting of the parts into a whole. It is the stage of deciding what is the right pattern, acting on the decision, and monitoring the outcome of that decision. Activity on this, fourth stage should never be fully computerized; rather it should be computer-assisted.

An interesting windfall of the service here envisaged is that all records of one individual will then be regularly collated as a mere routine activity of the (almost) fully computerized (self-monitoring) part. Patients hard to diagnose tend to switch physicians. The collating of their records may alert the next physician consulted with the information that here is a hard-to-diagnose patient--due to a repeated misreading of an item such as a rare disease, malingering, or any other cause. Pattern-recognition procedures alter when new background information is added to the effect that the pattern to be identified is hard to identify.

3. COMPUTERIZED DIAGNOSTIC SERVICE TODAY

How should the comprehensive computer-assisted medical diagnostic service grow out of the existing partial services? A lot of work has been done the world all over in the effort to implement partial computer-assisted diagnostic services, usually in those domains of medical diagnosis where cost effectiveness is most inviting. Diagnostic expert-systems services have been quite successfully implemented in those restricted and well-defined domains of medical diagnosis, not only as means for facilitating diagnostic tests, such as the magnificent instruments of computerized axial tomography, but also as a means for facilitating the process of pattern recognition. Facilitating the process of pattern recognition constitutes an integral part of the interpretation of diagnostic data, especially in the domains of electrocardiography, congenital heart diseases, and even in the diagnosis and treatment of certain infectious diseases. Thus, expert systems are to be found not only in computer software and in (self-)diagnostic services; they are these days already also built in some very powerful hardware. (The distinction between hardware and software is not hard-and-fast.) Moreover, computer banks have been used to stock data concerning markers for genetic diseases which are already rendering a highly important service to public health planning; likewise they may be used for stocking of genetic data for potential rare donors for blood transfusion and all sorts of transplantations.

Each of these services, being partial, is an approximation to a part of the recommended comprehensive computer-assisted medical diagnostic service. Moreover, each attempt at a computerization of a diagnostic service should be checked as an approximation to the best available physician's performance and to the one to be expected of the comprehensive service. Most methods used for the enlisting of computers in the aid of diagnostic services aim at some ambitious goal: they are intended to be expert systems, simulating the performance of top expert physicians. These methods have been concerned mainly with the stage of the pattern-recognition process focusing on the selection of guidelines for the assignment of specific diseases to individuals. In addition, operating already in the market are interactive user-computer systems, amenable to readily available user-friendly programming techniques.

This can be all to the good, and with proper monitoring it will be. Any attempt to extend the existing practices would follow one of two possible routes. The first is the construction of a broader system, on the way to a comprehensive, well-monitored service. The second is the improvement of the performance of the aids to pattern recognition (as aids to diagnosis), on the way to computer simulation of the whole diagnostic process (free of all human intervention except for the monitor). We prefer the first option on the grounds of both ethical and cost-effectiveness considerations. For the time being, however, the aim of a partial diagnostic expert-systems service is to help diagnosticians sort out patients into strictly defined classes of diseases (where some classes include exceptionally ambiguous cases, of course). How well a service can do this, and how usefully, is its first measure of efficient performance and overall desirability.

Traditionally there are three methods of or clusters of techniques for formalizing diagnosis:
(1) ignore all noise;
(2) allow for noise by introducing alternative options and giving each a statistical weight;
(3) utilize computer simulations, Artificial Intelligence or expert systems proper.

This list of three methods or clusters of techniques is not exhaustive: other successful methods or techniques are currently in use. These three methods or clusters of techniques will be described here only briefly--in order to contrast them with our proposal, the proposal for the immediate though partial implementation of the comprehensive computer-assisted medical diagnostic service.

Our interest in the methods of computer-assisted diagnosis currently in use is the following. Though the designers of the present successful partially-computerized diagnostic services employ the systems-analytic approach, they may complicate matters by the assumption that they employ the above three methods or clusters of techniques, individually or in combination. This will not totally hamper progress, because of a certain asymmetry. The systems analysis approach is open to any and all of the above three methods or clusters of techniques, each within limits. These methods or clusters of techniques, however, do not necessarily combine with

the systems analysis approach to medical diagnosis. If and
when they do, they seldom pertain to the system of
diagnostics--though they haphazardly relate to it: diagnostics
(or diagnosis as such) is rarely discussed or studied. Hopefully
the rise of interest in computer and systems analysis may help
sensitize writers to diagnostics and allow the system-analysis
approach to exert a full impact on the field of medical
diagnosis, namely, the prescribing of reasonably well-defined
functions and limits, subject to comprehensive cost-effectiveness
computations. But the systems approach does not depend on
the computer; on the contrary, it may help decide the extent
of the use of computers, in diagnosis as elsewhere.

    [Not all formalization is profitable. Is all knowledge
formalizable at least in principle? The answer is, definitely no.
Not only the growth of science and of language makes it
impossible because it is logically impossible to predict the future
growth of knowledge, but even the possibility of growth, when
embedded in language, makes full formalization impossible.
Only a dead language is fully formalizable. (At worst, rules for
each single existing sentence may have to be written down;
and with a dead language this is, perhaps, an exhaustable
task.) In this regard, medical competence (and its roots in
intuition) is no different from linguistic competence, and what
can be said about medical competence had been said about
linguistic competence by Noam Chomsky and by Yehoshua Bar-
Hillel in the early days of computers. Yet well defined chunks
of language are fully, even easily, formalizable to varying
degrees of success.]

Ignoring all Noise:

The first method or cluster of techniques of formalizing, is
known as the method of decision trees, or of tree-generation.
This method is currently widely employed in exercises given to
students in various schools, e.g., business schools and medical
schools. It resembles directions a stranger receives in a city:
next junction go right, straight, or left, and so on. Each time,
however, the instruction is conditioned upon the terrain: right
to the next junction and, if traffic is heavy (or rather, probably
heavy), then go right again, and so on. Moreover, information
might and should be included concerning the various (certain or

probable) expenses (such as toll or damage to tires) incurred at the junction. In diagnosis this is the strict technique of differential diagnosis with a clear-cut differential or crucial test at each junction plus the costs of these crucial tests, be it the cost in terms of money, time, or the health risk entailed.

Reliable decision trees should be put to use when available at reasonable cost. When reasonably available, or when unavailable except at exorbitant cost, they are unproblematic. In between, system-analytic considerations may add estimates. The cost of the consideration itself might also enter the consideration, and it will be low enough once a comprehensive computer-assisted medical diagnostic service will be in practice. For example, cheap differential decision trees apply to the standard urine tests that are available in standard clinics. Unavailable ones are sought these days as means for distinguishing between different causes of neuropsychiatric diseases. Such techniques exist, yet they are very partial and unclear. A borderline case is any unclear method, especially when costly.

Crucial tests, which are parts of decision trees, may be extremely helpful when feasible. They are as rational as possible, even when based on false theory since they clearly define moves at clear junctions and so afford (1) quicker attempts at correcting errors or (2) progressing with the diagnosis as far as reasonable or (3) the recognition of the limits of current knowledge (and the subsequent call for research). Yet they are neither sufficient nor always advisable. The information processing procedure in medical diagnosis may be much too complex to be formally represented by effective decision trees. In such cases attempts are often made to rectify the deficiency of the decision tree method by either the superposition on it of another method or cluster of techniques-- to be presented next--or by more commonsense considerations of cost effectiveness.

Noise plus probability:

The second method or cluster of techniques is statistical and rests chiefly on Bayes' theorem. That theorem allows to invert probabilities of cause and effect. Given a complete list of possible causes and their likely effects, it is not easy to conclude the cause of any given effect, since different causes (etiologies, diseases) can generate the same effect (disease, syndrome, symptom). When all the degrees of probability, of any given cause from a complete list of causes leading to any given effect from a complete list of effects are known, the theorem is a means to calculate the chance of determining the true cause for some given effect. The conditions under which the theorem holds are excessively stringent and they seldom obtain. It is a logical fact that this method can only be used by a comprehensive system. The theorem applies given all the possible causes of an effect, and given every probability measure connecting each cause with each possible effect--which is rarely the case.

In practice, physicians simply surmise most of these connections; the outcome is thus quite questionable. This is not to say that Bayes' theorem can be of help in daily diagnostic practice, because even when the conditions for Bayes' theorem hold, the situation may be very problematic and the theorem may offer no help. For example, the knowledge that abdominal pain hints at an appendicitis with a considerable yet low probability, does not solve the diagnostic problem; it is only a clear statement of the problem. Statistical method, even when most applicable, is in itself scarcely more than a merely marginal aid--however laudable--to proper diagnosis: rather, statistics can and should help decide whether and which crucial test is best to undertake at each diagnostic juncture.

The partiality of decision trees and statistical methods is not only in that they are confined to given classes of diseases and are partial techniques for diagnosing them. The order which the diagnostician picks the details may significantly influence the final diagnosis, given the same set of complaints, or even symptoms, and given a set of diseases or etiologies, and method. This certainly goes contrary to common sense. Here computerizing may help, in that its program should be order-insensitive. Yet the cost of such a procedure amounts to the cost of doing the series of tests in many possible orders: it may thus be quite prohibitive. This is also no news; students of computer diagnostic techniques have attempted to overcome this obstacle by empirical means: they observed that the best specialist clinicians operate in their diagnostic capacities, in order to emulate their series of examining complaints, signs and symptoms and possible diagnoses. This will not do (the technique is not sufficiently robust): observations will revalidate the knowledge that good diagnosticians are flexible enough not to stick to one routine. Can a computer be taught flexibility and inventiveness? Hardly.

Computer Simulation:

In medical diagnosis, computer simulation techniques were highly welcomed, perhaps because computerized diagnosis aimed at the expert's standard and with the aid of this approach ordinary physicians had hoped to be able to emulate the artist's intuition at work. Some measure of success is claimed for this method. By observing good physicians, reputed for their success in the clinic, some practical guidelines which they often used were formulated and skillfully fed into the computer as part of computerized diagnostic programs. In principle this is not different from watching a diagnostic video-tape of an expert at the bench. Likewise, some success was claimed for the replacement of some intuitions with quickly computable results. This is to be expected, but only in measure.

Uncertain predictions and judgments are seldom made in accord with the calculus of chance or the statistical theory of prediction. Instead, people often rely on a few rules of thumb which sometimes yield reasonable judgments, but sometimes lead to severe systematic errors. Tversky and Kahneman have shown that people often select or order outcomes by the degree to which the outcomes represent the essential features of the evidence, that is, by pattern recognition based upon the claim that a given group of phenomena is representative--whether by symptoms and signs of a known pattern, syndrome, disease, or etiology. In many situations such methods quickly render some outcome more likely than others. Since factors like prior probabilities of outcomes and the reliability of the evidence--all of which effect the likelihood of the outcomes--are often ignored when the rules of thumb are used, judgment based on them is often erroneous, and even systematically so. When several interacting factors intertwine in the evaluation of the probability of complex events, only the simplest and most available portions of past experience are likely to be considered. (This is known in the professional jargon as the compelling-scenario paradigm of decision making.) In this way images of the future, as well as the development of differential decision making procedures, might be easily distorted by scanty past experience. All this amounts to the familiar if lamentable inability to imagine anything but what has been observed. This is malignant in medical diagnosis as elsewhere, since it is self-reinforcing and misleading. Yet at times experts in the clinic cannot but justify a tentative diagnosis by reference to current statistics or to some previously encountered odd pattern.

What is to be done?

Since computer simulation is thus far so difficult and so inefficient, attempts were made to enrich it by letting the simulating computer beat the intelligence it imitates on matters on which computers generally have advantage over humans, i.e., in their ability to exploit both huge memory banks and fast calculations. Hence the limited but effective computer-assisted expert-system software programs now in the market. The complications caused by the use of vast data banks, however, calls for the aid of computer data retrieval techniques. The better the system, the more costly it is; and the more costly, the less available it is--at least prior to proper cost effectiveness.

Noise Management:

Each of the three existing methods has its use. Yet they are used tentatively since they are not placed in their proper contexts. They all suffer from an immense limitation of the knowledge of the domains of their validity and of the validity of the techniques incorporated in any or all of them.

Computer simulation methods are salutary since the computer can easily run the live diagnostician's gauntlet much faster and, if need be, announce in advance the sad end of a road. In particular, the computer simulation part of the computer system may clash with its ability to compute so well. Consider the case where the emulated human may skip a step in the calculation with the aid of a hypothesis, whereas the computer may get stuck. If it is taught to emulate humans and use both calculations and hypotheses, the computer may at times end with a contradiction. The diagnosticians emulated

can and should learn from this. At least, at times they are
known to have done so, and some of them have even learned
to improve upon the computer and thereby on their own earlier
diagnostic technique. Clinical diagnosis and instrumental
diagnosis thus intertwine. This fact is relevant to
considerations concerning the standards of responsible
implementation of expert-system software program.

Computer simulation is expensive: the cost-effectiveness
of a simulation should determine when its use is efficient
enough. Otherwise human-assisted computations should be
used. Whole branches of computation have been opened by
proving that certain calculations which are mathematically
feasible but technically unfeasible can be rendered feasible by
the omission of certain steps in the calculation and their
replacement by sheer guesses can be checked later on. Even
the purest mathematical uses of computers are thus improvable
by human intervention to the extent that certain well aimed
slight human interventions (which may be programmable at
times) bring practically inaccessible computations well within
reach.

To recapitulate, three methods of formalizing diagnosis
are at hand: the first, ignoring all noise, is unrealistic; the
second, allowing too much noise, has a limited applicability;
and the third, though aiming at emulating the expert, cannot
guarantee success and has also proved problematic. Systems
analysis permits combining them as efficiently as possible and
within the large context of diagnostics where humans and
machines interact.

4. THE FUTURE OF DIAGNOSTICS:
THE PLACE OF THE COMPUTER IN IT

Diagnostics has never gained the attention it deserves, and
there is almost no literature on it, even though the medical
literature is, to a large extent, applied diagnostics.
Nevertheless, no one doubts that it is, strictly speaking, a
science no more than a century-and-a-half old, and one which
has progressed magnificently. Diagnostic tools, hard and soft,
are breathtaking. Nosology, without which all such tools would
be useless, has advanced tremendously in the last century, and,
more so, in the last decade. Nevertheless, when the earlier

versions of this book were drafted about one decade ago, most experts consulted were clearly and unhesitatingly discouraging. This initial discouragement perhaps was directed at the proposal for a change in medical lifestyle that this book advocates--in addition to the more obvious hostility to our preference for computer assisted diagnosis over fully computerized one at a time of over-confidence in fully automated Artificial Intelligence.

What is required of a diagnosis is that it should be quick, cheap, and yielding accurate results. This can be done only to some degree; nonetheless, it may be hoped that adequate diagnosis may be achieved at an even lower cost with ever increasing speed and accuracy. Error would become increasingly less frequent and less harmful; and standards of health will hopefully increase as well--due to progress in both diagnosis and treatment, and to the improvement of the quality of life in general. All this is conditioned on improved precision. Today this is already unthinkable without the use of the computer--in medical research, in diagnostic and therapeutic activity, in medical administration and public health, and of course, in the very diagnostic process.

It is generally accepted that changing lifestyles as a method of treatment is preferable to that of aggressive intervention: physicians regularly recommend this to heart patients, for example and to patients suffering from cardiovascular diseases, hypertension and migraine. The integration of the computer in modern society (the modern medical establishments included), is it a medically commendable change of lifestyle? Clearly, the very existence of addiction to computers indicates that not all aspects of this change of the modern lifestyle are to the good. If one agrees that some diseases are in part by-products of lifestyles, then their diagnosis includes the study of our lifestyle. By the same token, a study of the place of computers in lifestyles would count as a part of modern epidemiology.

This discussion shows that diagnostics is open-ended, that ultimately diagnostics must embrace the study of individual and society at large, of social and political life and lifestyles at large and of the place of illness in them (a point familiar to readers of Samuel Butler's classic *Erewhon*). One can take science as a whole as a major factor determining the lifestyles of all the many different industrial and post-industrial

societies. One may even take one aspect of science in society--
the fascination with science--which determines the style and
prevalence of hypochondria and of addiction. It even
determines the addiction to science as expressed in the peculiar
forms of hypochondria, pill-popping, harmful physical exercises
and diets, addiction to computers, penny-arcades, and some
forms of research which are *a priori* futile. The aggressive
aspects of these non-aggressive alternative lifestyles indicate that
they are forms of addiction which blur the boundaries between
the non-aggressive and the aggressive and between healthy
curiosity and its obsessive subversion.

What produces fascination with an addiction to science
and glittering technology? It is with the promise of amplifying
human capacities and their projection beyond any given
recognized limit. The fascination and the addiction thus may
betray despair to be alleviated by magic. This may undermine
the possibility of improvement--the improvement sought being
that of human capacities and to as great an extent as possible.
For this the help of scientific and non-scientific technology has
to be enlisted, systematically, carefully and non-aggressively.

The initial incentive to write this book was the distaste
for the fascination with the computer to the extent of wishing
to relegate to the computer all sorts of responsibilities,
including the tasks of the diagnostician. Fortunately, by now,
in the field of computer-assisted medical diagnosis the
fascination is largely under control. This is mostly due to a
more practical appreciation of the computer's capacities and
limitations. This book advocates concern for practical
calculations but it also recommends restrictions to be placed on
computers in medical diagnosis and elsewhere. Unless these are
recognized and brought to bear on diagnostics in general and
the computer revolution in medical practice in particular one
cannot hope that realistic curiosity is under critical check.

Computers are now fashionable and they have entered
the heart of medicine during a period in which a few diagnostic
expert-systems software programs proved useful in a limited
way, and, at times, they proved more prudent than human
physicians (for instance, in the case of detection of suspected
digitalis toxicity). A few years ago the few Artificial
Intelligence researchers and their small crowd of followers
dreamt of replacing live physicians with computers. The dream

was met with general suspicion and hostility. An early draft of this book succeeded to arouse enormous hostility among publishers' readers, regardless of whether they were supporters of an Artificial Intelligence takeover or opponents to any intrusion of the computer into the clinic--because in that draft both views were rejected in favor of a comprehensive computer-assisted medical diagnostic service. Ever since the (rather humble) success of a few commercially available expert-system programs received recognition, the tide turned. A summary of the draft of this book appeared then and received some attention: it has appeared twice in English and once in German. Today the literature regularly refers to the 1982 paper by Zyporin as a classic since it presents as *passé* both the rejection of the computer and the acceptance of the computer as a surrogate diagnostician; instead, it advocates the use of the computer as an aid to human diagnosis. Zyporin's view is now sanctified even by Artificial Intelligence leading experts, notably Shortliffe and his co-workers. All this scarcely amounts to a daring attitude: evidence reveals that even in the matter of the maintenance and repair of commercial aircraft engines, something incomparably easier to formalize than diagnosis and treatment of humans, the interactive computer-assisted expert system is much superior to the fully computerized one.

Computers are now fashionable. Any professional body seeking added intellectual glamor and status may rent computer-time and hire a computer programmer. This is no substitute for considering the cost effectiveness of proposing cheap means of increasing the effectiveness of computer methods. In general, the most efficient means of contributing to the improvement of matters in the field of medicine at large is the recommendation to create a decentralized diagnostic computer network centrally coordinated with the aid of a central bank, to assist and not replace live diagnostic services. This plan is no utopia: it is utterly practical and rests on individualist ethics that forbids the relegation of responsibility to a machine while advocating the use of the machine, especially in order to save lives. It is particularly imperative to examine our factual considerations which present the rendering current computer services at once parts of a future comprehensive computer-assisted medical diagnostic service: we contend that this is a reasonable immediate guideline for planning rationally the next steps of the diagnostic services, thus forging the way to an intermediate future, in which the project will be under way.

Can one ignore the intermediate goal and center on the immediate goal alone? The comprehensive system is proposed here not as the immediate goal, but as the medium-range goal, even as a means to determine immediate goals. Strange as it may sound to one not yet versed enough in systems analysis techniques, even without the light of cost-effectiveness considerations, it is apparent that constructing a comprehensive computer-assisted medical diagnostic service should not be rushed: it would be both unwise and irresponsible to start at the very end, or perhaps very utopian. Yet not positing intermediate goals makes all short-range goals too arbitrary.

It is no accident that in the vast and rapidly growing literature of the social sciences there is no discussion of medium-range proposals. Long-range proposals are utopian; their implementation is not required and they are used as regulative principles, as metaphors for our deepest values. Usually actions are analyzed as adequate for immediate goals, whose character is allegedly influenced by the long-range ones. In the medium-range, ends and mean interact: ends previously unfeasible (and thus possibly utopian) may become feasible, barely feasible to some degree, depending on the growth of means. All the scientific and technological projects which deal with more than mere gadgetries can serve as examples for this; so can the constructions of technologies which are more than mere improvements on existing gadgets and whose applicability is still not clear and which serve new goals, such as the creation of new lifestyles, or the significant diversification of existing ones. New lifestyles, as I. C. Jarvie has stressed, often and unintentionally bring about new values and new goals.

The example at hand, the comprehensive computer-assisted medical diagnostic system, is such a medium-range goal; the contemplation of which may develop new goals and new tastes today.

The reasons in favor of the comprehensive computer-assisted medical diagnostic system are very obvious and compelling. Equipped with the techniques of thinking in terms of approximations, it is easier to begin with some idealization, while leaving at hand fairly useful approximations to it. The idealization may help in drawing a blueprint--the way the ideal geometrical figure helps drafting a real figure: both are regulative ideas. The regulative idea here is that diagnostics at large is the context of any computer application to medical diagnosis; the computer is the current context for any study of and improvement in diagnostics--in theory as well as in practice--due to its stress on precision, formalization, ensemble-representation and nodal points of decision for the sake of control, and due to its forceful and inevitable presence.

While building partial systems it is very useful to ask, Can they integrate, and if so how? Systems do regularly merge. Therefore, when designing a system it is fairly easy to avoid pitfalls that might be serious obstacles on the way to growth and/or integration. Some pitfalls are well known to designers, who regularly avoid them; others are not. These might be worse handicaps in medical diagnosis than elsewhere, particularly as the accumulation of dead weight of useless computer programs and data, and as the inflexibility of the theoretical apparatus increases.

It is thus easier to begin with the outline of a comprehensive computer-assisted medical diagnostic service and then design partial ones in more detail. After the idea of an overall service is made clear, those parts of the partial system that may increase value in integration may serve as incentives to growth. Thus, for example, data that have to be disposed of may first be statistically processed, so that the reduced net outcome be useful for future interactions with other partial systems leading to integration.

In abstract terms, the idea of a comprehensive system is an inducement to invest efforts in careful analysis and rational planning of such a system, in order to prevent unnecessary future obstacles to it. This way of looking at things makes any set of partial systems a system out of equilibrium, where the comprehensive system is the equilibrium point towards which the market and progress should force the partial systems. One major factor in the comprehensive system is the dissemination of information; firms selling partial services cannot avoid disseminating some information, and they may discover that it is to their advantage to offer an increasingly wider choice of information to their clients to purchase. This, too will render the present system moving towards the comprehensive system as an equilibrium point.

Some forces take the system to the opposite direction, the forces of resistance, even within the medical profession, to any computerization of any medical service, administrative, diagnostic, technical, or therapeutic. At times opposition may be overlooked. Physicians using a lab service--for any purpose--seldom inquire into the mechanics of that lab's working; and so they do not object to the tendency of such labs to automatize parts of their operations. The same holds for the administration of all medical services. In a private or cooperative clinic a practitioner may oppose the automation of the administrative services. The members of large medical organizations have neither the knowledge nor the authority to prevent it, and it takes place--at times excessively.

Generally speaking, as long as unchecked excessive automation is possible, opposition to it may very well be valid. The highest cost benefit should accrue from legislation of controls against excessive automation and from legislation leading to improved education. Education is particularly needed for the enhancement of both the ability to use automata and the ability to study and debate proposals for the reasonable limitation on them.

Surpassing the limits of automation is necessarily expensive and possibly dangerous. The absence of criteria for when it is useful and laudable to implement computer services, especially in the service of diagnosis, explains as rational the expression of fears and of reluctance which some large institutions display, concerning the implementation of existing programs. This book offers a context in which to consider such fears as legitimate and enlist them to the aid of furthering understanding of their general context which is diagnostics in general. Within this most general context, simple and commonsense rules can be implemented soon to limit the applicability of computers and to institute simple means for the monitoring and testing of that applicability and for reversing or modifying erroneous decisions. This could facilitate debate in the medical institutions that resist the implementation of computer services. Designers of such services may better interact with these medical institutions and with prospective users of those services there, as co-designers and as spokespeople for customers--be these patients or physicians. The interaction should bring mutual influences. That is to say, designers and users--programmers, patients and physicians--may all benefit from detailed debates about computerization. This book offers means for beginning these debates. The sooner this book will be obsolete the happier its authors will be.

Large medical concerns--from hospitals and clinic chains to insurance companies and the pharmaceutic industry--are now increasingly aware of the increasing need to consider the place of the cost of medical services in all sorts of budgetary considerations. Books bring the issue to public awareness. This brings up both hope and threat: bureaucrats should not have all the power of decision making. The comprehensive computer-assisted medical diagnostic service here recommended may help reduce the threat. The computer part of this diagnostic service should be open to the public for examination and study, with private cumulative records restricted to their individual owners, thereby increasing the freedom of both physician and patient. It should be a much more physician-oriented and much more patient-oriented diagnostic service than now available. The available service is extremely limited and is couched in a paternalist language. Paternalism fosters a monopoly and it is not in the best interest of physicians to monopolize their patients. As Adam Smith observed centuries ago, monopoly goes not only against the public interest but also against the better self-interest of the monopolists themselves. Yet physicians need to be enlightened as to the economic aspect of their profession no less than as to the foundations of their practice in both science and morality-- otherwise they could not partake in any meaningful reform of it; this demands a broader framework than the current adversarial framework that serves currently as the exclusive forum for discussions of medical ethics and medical administration. The current framework ought to be transcended: it has outlived its usefulness.

The informed consent of the patient is a matter of patient participation in all medical decision making to degrees much higher than are now in evidence, which require much higher levels of education of the ordinary citizen in matters medical than are now attained. (The same holds for the matter of the exercise of autonomy by the ordinary citizen.) More knowledge is required and more active participation; *knowledge without participation is lame; participation without knowledge is blind.* Politics these days increasingly develops along these lines as the demand is rising for both political education and for a pluralist participatory democracy.

The computer-assisted diagnostic service is a powerful means for educating the general public as well as the medical profession towards pluralism; for inviting individual patients to assume responsibility for their own treatment; for inviting medical diagnosticians to raise their critical attitude and awareness of current diagnostic theory and procedures and their readiness to review them critically with the hope of improving on them; for reducing the monopoly of physician over patients in diagnosis and treatment; and for considerably reducing the prevalent physician-patient friction and adversary relations. Saving lots of money on insurance and on litigation will be an enormous source of income for the further development of existing services furnished along the lines here described. (Insurance companies may find it profitable to invest in this service.)

Computers offer tremendous freedom, thereby facilitating the extension of services in quite innovative ways. For example, as patients are at the notorious mercy of two monopolies, the medical and the legal professions, it may be expected that sooner-or-later the service will include some legal-medical counselling, some forensic medical information. Since citizens' advice bureaus co-exist and since they are already helped out by computers, the very idea of a comprehensive computer-assisted citizens' advice service is not so remote from reality: what it needs is better theoretical background and some tightening up and coordination in accord with that knowledge.

There is no guarantee in matters educational. The computer has proven itself a tremendous aid for education, yet regrettably it has also created a new form of addiction; all addiction is loss of autonomy whereas the highest end of all education is autonomy. It is best to consider computer assisted diagnosis as a part of the means for facilitating patient-computer interaction: patients should consult physicians, peers, and lawyers. The role of properly instituted communication systems is to enhance all human interaction, not to replace it.

The ideal treatment is the non-aggressive, hardly intervening, "soft" prescription--the prescription which aims at a

restitution of software rather than a modification of hardware--
with low cost and high effectiveness. The treatment softest
and best of all is that which is limited to changes in lifestyle.
The following thought-experiment illustrates this. Two barely
distinguishable diseases will be distinguished with great
emphasis once it is found that one and only one of them is
given to "soft" treatment; such treatment will then tend to be
used as differential diagnosis.

Innovative medical technology invites diagnostic
techniques to insure the avoidance of the excessive use of
"hard" treatment; yet the use of diagnostic techniques itself
may become excessively "hard", including the "hard" use of the
computer both as a diagnostic tool and as a tool for the
diagnostic reasoning process. The computer should be used in
diagnosis, but gently.

Diagnostics in general should draw attention to a few
important items: the choice between different medical styles;
the possibility of improving health conditions by complementing
clinical diagnosis and instrumental diagnosis with each other as
well as by integrating diagnosis with monitoring and treatment;
the usefulness of evincing cost-effectiveness considerations
explicitly and in a critical attitude; and, finally, the worth of
communicating such considerations to patients and to the
public at large. Hopefully this will democratize medicine in
general and diagnosis in particular. Medicine can be a powerful
instructive tool and it may contribute significantly to continuing
progress of public education in both science and morality, and
thus to the process of democratization; it will then truly
integrate in the fabric of modern democratic society.

SELECTED BIBLIOGRAPHY FOR FURTHER READING

Technical items are marked with an asterisk.

Agassi, J., 1975, *Science in Flux*, Kluwer, Dordrecht.

Agassi, J., 1976, "Causality in Medicine", *J. Med. Phil.*, 1, 301-317.

Agassi, J., 1977, *Towards A Rational Philosophical Anthropology*, Kluwer, Dordrecht.

Agassi, J., 1978, "Liberal Forensic Medicine", *J. Med. Phil.*, 3, 226-241.

Agassi, J., 1979, "The Whole and Its Parts", *Nature and Systems*, 1, 32-36.

Agassi, J., 1980, "Between Science and Technology", *Phil. Sci.*, 47, 82-99.

Agassi, J., 1981, "Mechanistic and Holistic Models in Psychiatry", *Nature and Systems*, 3, 143-52.

Agassi, J., 1981, "Psychoanalysis as a Human Science: A Comment", *Brit. J. Med. Psychol..*, 54, 295-296.

Agassi, J., 1983, "Technology as both Art and Science", *Research in Philosophy and Technology*, 6, 55-63.

Agassi, J., 1984, "A Holomechanical Model for Research for the Life Sciences", *J. Biol. Stud.*, 7, 75-79.

236   SELECTED BIBLIOGRAPHY

Agassi, J., 1985, *Technology: Philosophical and Social Aspects*, Kluwer, Dordrecht.

Agassi, J., 1987, "The Uniqueness of Scientific Technology", *Mehtodology and Science*, 20, 1987, 8-24.

Agassi, J., 1988, "Analogies Hard and Soft", in D. Helman, Editor, *Analogical Reasoning*, Kluwer, Dordrecht.

Agassi, J., 1988, "Deadalos, Winter 1988", *SigArt Newsletter*, 105, July 1988, 15-22.

Agassi, J., 1990, "Democratizing Medicine", in G.L. Ormiston and R. Sassower, Editors, *Prescriptions: The Dissemination of Medical Authority*, Greenwood Press, Westport, CT.

Agassi, J., and Jarvie, I.C., 1987, *Rationality: The Critical View*, Kluwer, Dordrecht.

Agassi, J.B., 1979, *Women on the Job*, Lexington Books, Lexington MA.

*Aikens, J.S., 1983, "Prototypical Knowledge for Expert Systems", *Artificial Intelligence*, 20, 1983, 163-210.

Applebaum, P.S., Lidz, C.W. and Meisel, A., 1987, *Informed Consent: Legal Theory and Clinical Practice*, Oxford University Press, New York.

Ashby, W.R., 1956, *An Introduction to Cybernetics*, Chapman and Hall, London.

*Austin, C.J., 1988, *Information Systems for Health Service Administration.*, Health Admin. Press, Ann Arbor MI.

*Bailey, J.T.J. and Thompson, M., Editors, 1975, *System Aspects of Health Planning*, North-Holland Publn., Amsterdam.

Bare, A. and Feigenbaum, E.A., 1982, *Handbook of Artificial Intelligence* (Vols. 1 & 2), Kaufmann, Los Altos CA.

Bar-Hillel, Y., 1964, *Language and Information*, Addison Wesley, Reading MA.

Bergman, A.B. and Stamm, S.J., 1967, "The Morbidity of Cardiac Nondisease in School Children", *N. Engl. J. Med.*, 276, 1008-1013.

*Berki, S.E., 1972, *Hospital Economics*, Lexington Press, Lexington MA.

Boden, M.A., 1977, *Artificial Intelligence and Natural Man*, Basic Books, New York.

*Buchanan, B.G. and Shortliffe, E.H., 1985, *Rule-Based Expert Systems: The Mycin Experiment*, Addison Wesley, Reading MA.

Bulger, Roger J., Editor, 1986, *In Search of the Modern Hippocrates*, University of Iowa Press, Iowa C Ity IO.

*Bundy, A., Byrd, L. and Mellish, C.S., 1983, "CRSL: A Language for Expert Systems for Diagnosis", *Proc. IJCAI-83*, Karlsruhe, 1983, 218-221.

Bunge, M., 1979, *A World of Systems*, Kluwer, Dordrecht.

Bunge, M., 1981, "Analogy Between Systems", *Intl. J. General Systems*, 7, 221-230.

*Bunker, J.P., Barnes, B.A. and Mosteler, F., 1977, *Costs, Risks and Benefits of Surgery*, Oxford University Press, New York.

Cady, D., Editor, 1978, *Computer Techniques in Cardiology*, Academic Press, New York.

Chapman, L.J. and Chapman, J.P., 1969, "Illusory Correlation as An Obstacle to The Use of Valid Psychodiagnostic Signs", *J. Abnorm. Psychol.*, 74, 271-280.

238   SELECTED BIBLIOGRAPHY

Charniak, E., 1983, "The Bayesian Basis of Common Sense Medical Diagnosis", *Proc. AAAI-83*, Washington DC, 70-73.

Clancey, W.J., 1983, "The Epistemology of Rule-Based Expert Systems", *Artificial Intelligence*, 20, 215-251.

Corcoran, D.W.J., 1971, *Pattern Recognition*, Penguin, Harmondsworth.

*Couger, J.D. and Knapp, R.W., 1974, *System Analysis Techniques*, John Wiley, New York.

Crofts, D.J., 1972, "Is Computerized Diagnosis Possible?" *Comp. and Biomed. Res.*, 5, 351-367.

Dinstein Y., 1965, *The Defense of "Obedience to Superior Orders" in International Law*, Sijthof, Leyden.

Dombal, F.T., Leaper, D.J., Harocks, J.C., Stanisland, J.R. and McCann, A.P., 1974, "Human and Computer-Aided Diagnosis of Abdominal Pain: Further Report with Emphasis on Performance of Clinicians", *Brit. Med. J.*, 1, 376-380.

*Driggs, M.S., Editor, 1976, *Problem Directed and Medical Information Systems*, Intercontinental Medical Books, New York.

Dubos, R., 1959, *The Mirage of Health, Utopian Progress and Biological Change*, Anchor Books, New York.

Dubos, R., 1966, *Man and His Environment, Biomedical Knowledge and Social Action*, Pan-American Health Organization, Scientific Publication No. 131, Washington D.C.

Durbin, P. T., 1980, *A Guide to the Culture of Science, Technology and Medicine*, Free Press, New York.

Elling, R.H. and Sokolowski, M., Editors, 1978, *Medical Sociologists At Work*, Transaction Books, NJ.

Engelhardt, H.T. and Spicker, S., Editors, 1979, *Clinical Judgment*, Kluwer, Dordrecht.

Farvar, M.D. and Milton, J.P., 1972, *The Careless Technology*, Natural History Press, Garden City NJ.

Faust, D., 1986 "Research on Human Judgment and Its Application to Clinical Practice", *Professional Psychology: Research and Practice*, 17, 420-430.

Faust, D. and Miner, R.A., 1986, "The Empiricist's New Clothes: DSM-III in Perspective", *Am. J. Psychiat.*, 143, 962-967.

Faust, D..and Ziskin, J., 1988, "The Expert Witness in Psychology and Psychiatry", *Science*, July, 1988, Vol.241, 31-35.

Feinstein, A.R., 1967, *Clinical Judgment*, Krieger, Molabar FL.

Feinstein, A.R., 1987 *Clinimetrics*, Yale University Press, New Haven CT.

Finberg, H.V. and Hiatt, H.H., 1979, "Evaluation of Medical Practices, The Case for Technology Assessment", *N. Engl. J. Med.*, 301, 1086-1091.

Fletcher, J., 1954, 1960, *Morals and Medicine: Moral Problems of The Patient's Right to Know The Truth etc.*, Beacon Press, Boston MA.

Foreman, J., 1989, "Computers in Clinical Medicine Raise Questions of Liability", *Arch. Ophthal.*, 107, 25.

Fried, Y. and Agassi, J., 1976, *Paranoia: A Study in Diagnosis*, Kluwer, Dordrecht.

Fried, Y. and Agassi, J., 1981, *Psychiatry as Medicine*, Kluwer, Dordrecht.

Friedman, R.B. and Gustafson, D.H., 1977, "Computers in Clinical Medicine, A Critical Review", *Comp. Biomed. Res.*, 10, 199-204.

Friedson, E., 1971, *Profession of Medicine: A Study of The Sociology of Applied Knowledge*, Dodds, Mead, New York.

*Garrett, D., 1973, *Hospitals--A Systems Approach*, Auerbach, Philadelphia PA.

Glazer, N., 1971, "Paradoxes of Health Care", *Public Interest*, 22, 62-77.

Goode, W.J., 1967, "The Protection of The Inept", *Amer. Sociol. Rev.*, 32, 5-19.

*Grams, R.R., 1972, *Problem Solving and System Analysis and Medicine*, Thomas, Springfield IL.

Groves, J.E., 1978, "Taking Care of The Hateful Patient", *N. Engl. J. Med.*, 298, 883-887.

Gussow, Z., 1989, *Leprosy, Racism, and Public Health. Social Policy in Chronic Disease Control*, Westview, Bolder CO.

Harel D., 1987, *Algorithmics: The Spirit of Computing*, Addison-Wesley, Reading MA.

*Harel, D., 1987, "Statecharts: A Visual Formula for Complex Systems", *Sci. Comput. Prog.*, 8, 231-274.

Harris, R., 1969, *A Sacred Trust*, Penguine, Baltimore MD.

Hart, H.L.A., 1963, *Law, Liberty and Morality*, Stanford University Press, Stanford CA.

*Haugeland, J., Editor, 1981, *Mind Design: Philosophy, Psychology, and Artificial Intelligence*, MIT Press, Cambridge MA.

Herzhaft, G., 1969, "*L'effet Nocebo*", *Encephale*, 58, 486-503.

*Hobbs, J.R. and Moore, R.C., Editors, 1985, *Formal Theories of the Commonsense World*, Ablex, Norwood NJ.

Hsiao, J.K., Bartko, J.J. and Potter, W.Z., 1989, "Diagnosing Diagnoses", *Arch. Gen. Psychiat.*, 46, 664-667.

Jarvie, I.C., 1972, *Concepts and Society*, Routledge, London.

Jarvie, I.C., 1985, *Thinking About Society*, Kluwer, Dordrecht.

Jones, Lovell A., Editor, 1989, *Minorities and Cancer*, Springer, New York.

Kaufmann, W., 1973, *Without Guilt and Justice: From Decidophobia to Autonomy*, Wyden, New York.

Kahneman, D., Slovic, P. and Tversky, A., 1982, *Judgment Under Uncertainty: Heuristics and Biases*, Cambridge University Press, New York.

Katz, J., 1984, *The Silent World of Doctor and Patient*, Free Press, New York.

Keele, R.F., 1988, "What is a Case?" *Arch. Gen. Psychiat.*, 45, 374-376.

Kisch, A.I. and Reader, L.G., 1969, "Client Evaluation of Physician Performance", *J. Health and Soc. Behav.*, 10, 51-58.

Kiefer, T. and Hansen, T., 1984, *Get Connected: A Guide to Telecommunication*, Culver City CA.

Kleinmunz, B., 1984, "Diagnostic Problem Solving by Computer: A Historical Review and the Current State of the Science", *Comput. Biol. Med.*, 14, 255-70.

Kunz, J.C., Shortliffe, E.H., Buchanan, B.G. and Feigenbaum, E.A., 1984, "Computer-Assisted Decision Making in Medicine", *J. Med. Phil.* 9, 135-160.

242   SELECTED BIBLIOGRAPHY

Laor, N., 1984, "The Autonomy of the Mentally Ill: A Case
    Study in Individualistic Ethics", *Phil. Soc. Sci.*, 14,
    331-349.

Laor, N., 1985, "Prometheus the Imposter", *Brit. Med. J.*, 290,
    681-684.

Laor, N., 1985, "Psychoanalysis as Science: The Inductivist's
    Resistance Revisited", *J. Amer. Psychoanal. Assoc.*, 33,
    149-166.

Laor, N., 1987, "The Poverty of Current Forensic Psychiatry",
    Phil. Soc. Sci., 17. 571-578.

Laor, N., 1987, "Toward Liberal Guidelines for Clinical
    Research with Children", *Law and Medicine*, 6, 127-137.

Laor, N., 1990, "Seduction in Tongues: Reconstructing the
    Field of Metaphor in the Treatment of Schizophrenia", in
    G.L. Ormiston and R. Sassower, Editors, *Prescriptions:
    The Dissemination of Medical Authority*, Greenwood Press,
    Westport, CT.

Leach, G., 1970, *The Biocrats: Implications of Medical
    Progress,* revised edition, Penguin, Harmondsworth.

Lidz, C.W., Meisel, A., Zerubavel, E., Carter, M., Sestak, R.M.
    and Roth, L.H., 1984, *Informed Consent: A Study in
    Decisionmaking in Psychiatry*, Guilford Press, New York.

Lieff, J.D., 1987, *Computer Application in Psychiatry*, American
    Psychiatric Press, Washington DC.

*Lilienfeld, A.M., 1976, *Foundations of Epidemiology*, Oxford
    University Press, New York.

Lipsitt, D.R., 1980, "The Patient and The Record", *N. Eng. J.
    Med.*, 302, 167-168.

Manning, R.C., 1986, "Why Sherlock Holmes?", *Phil. Stud.*,
    117, 65-84.

*Martin, J., 1977, *Computer Data Base Organization*, Prentice-Hall, Garden City NJ.

McCleery, R.S., 1971, *One Life--One Physician: An Inquiry into The Medical Profession's Performance in Self-Regulation, A Report to the Center for the Study of Responsive Law*, Public Affairs Press, Washington D.C.

McCorduck, P., 1988, "Artificial Intelligence: an Apperçu", *Daedalus*, 117, 65-84.

Meador, C., 1965, "The Art and Science of Nondisease", *N. Engl. J. Med.*, 272, 92-95.

Meister, D., 1976, *Behavioral Foundation of System Development*, John Wiley, New York.

Miller, R.A., Schaffner, K.F. and Meisel, A., 1985, "Ethical and Legal Issues Related to the Use of Computer Programs in Clinical Medicine", *Ann. Intern. Med.*, 102, 529-536.

*Minsky, M., 1975, "A Framework for Representing Knowledge". Reprinted in Haugeland, J., Editor, 1981, *Mind Design: Philosophy, Psychology, and Artificial Intelligence*, MIT Press, Cambridge MA.

Mintz, M., 1965, 1967, *By Prescription Only: A Report on The Role of The United States Food and Drug Administration, The American Medical Association, Pharmaceutical Manufacturers, and Others in Connection with The Irrational and Massive Use of Prescription Drugs That May Be Worthless, Injurious, or Even Lethal*, Beacon Press, Boston MA.

Moor, P.G., 1968, *Basic Operational Research*, Pitman and Sons, London.

Munson, R.G., 1986, *Concepts and Design*, Prentice Hall, Englewood Cliffs NJ.

244   SELECTED BIBLIOGRAPHY

G.L. Ormiston and R. Sassower, Editors, *Prescriptions: The Dissemination of Medical Authority*, Greenwood Press, Westport, CT.

Patli, R.S., Szolowits, P. and Schwartz, W.B., 1981, "Causal Understanding of Patient Illness in Medical Diagnosis", *Proc. IJCAI-81*, Vancouver, 1981, 893-899.

*Patrick, E.A., Stelmack, F.P. and Shen, N.Y.L., 1974, "Review of Pattern Recognition in Medical Diagnosis and Consulting Relative to A New System Model", *IEEE Trans. on Syst. Manag. and Cyber.*, SMC-4, 1-16.

*Pauker, S.G., Gorry, G.A., Kassirer, J.P. and Shwartz, W.B., 1976, "Towards The Simulation of Clinical Cognition, Taking A Present Illness by Computers", *Amer. J. Med.*, 60, 981-996.

Pellegrino, E.D., 1986, "Rationing Health Care: The Ethics of Medical Gatekeeping", *J. Contemp. Health Care Policy*, 2, 23-45.

Pollak, V.E., 1983, "The Computer in Medicine", *J.A.M.A.*, 253, 62-68.

Popper, K.R., 1935, 1959, *The Logic of Scientific Discovery*, Hutchinson, London.

Popper, K.R., 1945, *The Open Society and Its Enemies*, Routledge, London.

Ranshoff, D.F. and Feinstein, A.R., 1978, "Problems of Spectrum and Bias in Evaluating The Efficacy of Diagnostic Tests", *N. Engl. J. Med.*, 299, 926-930.

*Reggia, J.A. and Tuhrim, S., Editors, 1985, *Computer Assisted Medical Decision Making*, Vols 1 & 2, Springer, New York and Berlin.

Rennel, G. and Shortliffe, E.H., 1987, "Advanced Computing for Medicine", *Sci. Amer.*, 257, 154-161.

Roberts, W.C., 1978, "The Autopsy: Its Decline and Suggestion for its Revival", *N.Engl. J. Med.*, 299, 332-337.

Russell, Louise B., 1989, "Some of the Tough Decisions Required by a National Health Plan", *Science*, 246, Nov. 17, 1989, 892-896.

Shaw, Bernard, 1911, 1946, *The Doctor's Dilemma*, Penguin, Harmondsworth.

Sheff, T.J., 1963, "Decision Rules, Types of Errors, and Their Consequences in Medical Diagnosis", *Behav. Sci.*, 8, 97-107.

*Shortliffe, E.H., 1976, *MYCIN: Computer-based Medical Consultation*, American Elsevir, New York.

Shortliffe, E.H. and Buchanan, B.G., 1975, "A Model of Inexact Reasoning in Medicine", *Math. Biosci.*, 23, 351-379.

Sidel, V.W., 1976, "Quality for Whom? Effects of Professional Responsibility for Quality of Health Care on Equity", *Bull N.Y. Acad. Med.*, 52, 164-176.

Slaughter, F.G., 1950, 1961, *Semmelweiss: The Conqueror of Childbirth Fever*, Collier, New York.

Szasz, T.S. and Hollander, M.H., 1956, "A Contribution to The Philosophy of Medicine: The Basic Models of The Doctor-Patient Relationship", *Arch. Intern. Med.*, 97, 585-592.

Szasz, T.S., 1956, "Malingering: Diagnosis or Social Condemnation?" *Arch. Neurol. and Psychiat.*, 76, 432-443.

Szasz, T.S., 1963, *Law Liberty and Psychiatry: An Inquiry into The Social Use of Mental Health Practices*, Macmillan, New York.

Szasz, T.S., 1980, *Sex by Prescription*, Penguin, Dallas PA.

Tischler, G.L., Editor, 1987, *Diagnosis and Classification in Psyciatry: A Critical Approach to DSM-III*, Cambridge University Press, New York.

Tomlinson, T. and Brody, H., 1988, "Ethics and
    Communication in Do-Not-Resuscitate Orders", *N. Eng. J.
    Med.* 318, 43-46.

Tullock, G., 1965, *The Politics of Bureaucracy*, Public Affairs
    Press, Washington D.C.

Tversky, A. and Kahneman, D., 1974, "Judgement Under
    Uncertainty: Heuristics and Biases", *Science*, 185,
    1124-1131.

Valliant, G.E. and Schnurr, P., 1988, "What is a Case?"
    Arch. Gen. Psyciat., 45, 313-319.

*Van Egmond, J., de Vries Robbe, P.F. and Levy, A.H.,
    Editors, 1976, *Information Systems for Patient Care*, North-
    Holland Pbln., Amsterdam.

Veatch, R.M. and Brown R., 1976, *Ethics and Health Policy*,
    Ballinger, Cambridge MA.

Weed, L.L., 1969, *Medical Records, Medical Education, and
    Rational Care*, Yearbook of Medical Pbln., 1971.

White, K.L., Anderson, D.O., Kalimo, E., Kleczkowski, B.M.,
    Purola, T. and Vukmanovic, C., 1977, *Health Services:
    Concepts and Information for National Planning and
    Management*, Public Health papers, No. 67, WHO, Geneva.

Wiener, N., 1950, *The Human Use of Human Beings:
    Cybernetics and Society*, Houghton Mifflin, Boston MA.

Wiener, N., 1964, *God and Golem, Inc., A Comment on
    Certain Points Where Cybernetics Impinges on Religion*,
    MIT Press, Cambridge MA.

Wigertz, O., 1988, "Making Decisions Based on Fuzzy Medical
    Data--Can Expert Systems Help?" *Meth. Inform. Med.*, 25,
    59-61.

Williams, R., 1962, 1968, *Communication*, Chatto and Windus, London.

Young, F.A., 1987, "Validation of Medical Software: Present Policy of The Food and Drug Administration", *Ann. Int. Med.* 106, 628-629.

Zinberg, N.E. and Robertson, J.A., 1972, *Drugs and The Public*, Simon and Schuster, New York.

Ziporyn T., 1982, "Computer-Assisted Medical Decision-Making: Interest Growing", *J.A.M.A.*, 248, 913-918.

INDEX OF NAMES

Alzheimer, Alois ... 195, 201
Aristotle ... 96, 186, 212

Bacon, Francis ... 49
Bar-Hillel, Yehoshua ... 219
Bayes, Thomas ... 147-148, 220-221
Bernard, Claude ... 131
Bertalanffy, Ludwig ... 97, 103-104
Bichat, Marie F.-X. ... 174
Bohr, Niels ... 57
Broca, Paul ... 195
Bunge, Mario ... xi, 97, 190
Butler, Samuel ... 225

Cabot, Richard ... 84, 174
Charcot, Jean-Marie ... 194
Chomsky, Noam ... 219
Churchman, C.W. ... 107
Crohn, B.B. ... 178

Darwin, Charles ... 99
Dinstein, Yoram ... 74

Einstein, Albert ... xi, 3, 27, 84, 92, 99

Feinstein, Alvin ... 174, 186
Freud, Sigmund ... 195, 201

Gödel, Kurt ... 100

Hart, H.L.A., ... 74
Hippocrates ... 74-75, 78, 175, 180, 188, 192, 194
Holmes, Sherlock ... 159

Jackson, John Hughlins ... 195
Jarvie, Ian C., ... 228

Kahneman, Daniel ... 35-37, 151-152, 157, 222
Kant, Immanuel ... 185
Katz, Jay ... 77, 79
Koestler, Arthur ... 43
Kraepelin, Emil ... 195
Kuhn, Thomas S. ... 88

Lakatos, Imre ... 204
Lidz, Charles W., ... 82
Lister, Joseph ... 127
Lyell, Charles ... 98

Manning, Rita ... 159
McCorduck, Pamela ... 160

Nobel, Alfred ... 160

Osler, William ... 84, 174

Pasteur, Louis ... 127, 160
Popper, Karl ... xi, 27, 78, 88, 99, 131, 185, 196

Quine, Willard V. ... 184

Semmelweis, Ignaz
    Philipp ... 127
Shannon, Claude ... 213
Shaw, Bernard ... 165
Shortliffe, Edward H. ... 227
Skinner, B.F. ... 189-190
Smith, Adam ... 231
Szasz, Thomas S. ... 142, 193,
    206-207

Tourette, George Gille ... 197,
    201
Tversky, Amos ... 35-37,
    151-152, 157, 222

Vineberg, A.M. ... 134
Voltaire, François
    Marie ... 165

Whewell, William ... xi, 185
Wiener, Norbert ... 204
Wigertz, Ove ... 186

Zadeh, Lofti ... 186
Zyporin, Terra ... 227

Abstraction ... 99, 123
Accidental symptoms ... 212
Action ... 51, 71-72, 77, 228
Addiction ... 225-226, 232
Administration ... xi, 83, 145
Aggression ... 178-179, 205,
    225-226, 232
AIDS ... 168, 196, 215
Ambiguity ... 218
Anamnesis ... 137, 213
Anatomy ... 3, 43
Anomaly ... 131, 215
Anti-psychiatry ... 206
Anti-social
    personality ... 197-198
Apprentice ... 65, 177
Approximation ... xi, 28, 92,
    99, 102, 109, 112, 119,
    153, 164, 166, 217, 228
Artificial Intelligence ... 41,
    49, 52, 64, 148, 150, 184,
    188, 218, 225-227
Authorities ... 11, 121
Authority ... 69, 79, 88, 109,
    174, 196, 230
Authorization ... 69-70
Autonomy ... 62-63, 71, 84,
    168, 181, 193, 231-232

Background ... ix, xi, 80, 148,
    190, 216, 232
Basic research ... 89-90

Bayes' theorem ... 147-148,
    220-221
Behavior ... 96-97, 106,
    189-190
Behavior modification ... 190
Beneficiaries ... 18, 58, 67-68,
    85-86, 89
Bias ... xiv, 43, 47, 50-51,
    94-95, 102, 183
Brittleness ... 42, 64, 149, 152
Broad-spectrum
    treatment ... 13
Budget ... 10, 14-18, 30, 45,
    164, 231
Budgetary constraints ... 16
Bureaucratization ... 64, 176,
    231
By-product ... 87, 89, 225

Check-up ... 62, 161, 163-164,
    169, 171-173
Chess ... 7, 86, 116
Choice ... xiv, 7, 13, 27, 37,
    42, 78, 106-108, 154, 211,
    215, 229, 233
Citizens' advice
    bureaus ... 232
Civil disobedience ... 74
Clinic ... ix, xii, 15, 26, 41,
    45, 51, 62, 67-68, 84-85,
    94, 117, 125, 181,
    220-224, 227, 230-231,
    233

Clinical diagnostic
    encounter ... 3, 33, 41,
    57, 60, 62, 65, 93, 117
Clinical ethical
    conference ... 175
Clinical practice ... 84, 125
Collectivism ... 68, 82, 105
Common sense ... xiv, 13-14,
    36, 62, 92, 108, 187, 191,
    200-201, 207-208, 214,
    220-221, 230
Communication ... 4, 87, 106,
    232
Compelling-scenario
    paradigm ... 222
Compensation ... 164
Complaint-orientation ... 165
Comprehensive computer-
    assisted diagnostic
    service ... xv, 1-2, 4, 12,
    78, 93, 111, 113, 143,
    145-146, 150, 161,
    163-164, 168, 173-176,
    193, 227
Compromise ... xiv, 78, 85,
    125
Computer literacy ... 57
Computer program ... 6, 9,
    11, 51, 62, 80, 139, 146,
    156, 213, 227, 229
Computer revolution ... 12,
    41-42, 95, 226
Computer simulation ... 25,
    115-116, 148-150, 218,
    222-224
Computer technology ... ix-x,
    6, 8, 11, 30, 42, 53, 83,
    92, 148
Computer-assisted diagnostic
    service ... xii, 5-6, 11,
    34, 39, 59, 61, 64-65, 71,
    84, 161

Computer-assisted diagnostic
    system ... 29, 60, 166
Computer-generated
    hypotheses ... 49
Computerization ... 31, 63-64,
    217, 230-231
Concealment of error ... 58,
    79, 123, 157-158, 167
Conceptual framework ... 67
Confidence factors ... 49
Confidentiality ... 16, 155, 168,
    175
Confirmation ... 2, 43, 118,
    159, 196
Conflict ... 13, 18-19, 37, 51,
    62, 73, 106-108, 125,
    165-166, 198
Conflict of interests ... 39,
    67-71, 75, 107-109, 111,
    130, 177
Consensus ... 9, 122-123, 125
Constraint ... 9-10, 16, 26, 29,
    71-72, 78, 103, 116
Consultation ... 54, 56, 64, 81,
    144, 157, 161-162, 164,
    170, 216, 225, 232
Consumers'
    organization ... 110-111
Context ... 18-19, 25, 27-31,
    33-35, 67-68, 70, 78, 85,
    95, 98-102, 107-108, 115,
    129, 149, 152, 163, 192,
    206, 210-211, 223-224,
    229-230
Contextualization ... 1, 19
Contract ... 69-70, 74-75,
    89-90
Contradiction ... 44, 51,
    100-102, 118, 148,
    191-192, 223

Control ... x, 12, 14, 18,
  30-31, 39, 47, 53, 56-58,
  60, 64, 66, 71, 76, 82,
  84, 88-89, 106, 109, 112,
  124, 134, 145, 147,
  155-156, 161-162, 168-169,
  174-177, 191, 214, 226,
  229-230
Control group ... 162
Controlled
  schizophrenia ... 43-45
Controversy ... 51, 53, 109,
  200-203
Cooperation ... 79, 103
Corporate individual ... 68, 72
Corroboration ... 51, 121, 160,
  208
Cosmetic operation ... 129,
  141-142
Cost benefit ... 7, 110, 113,
  143, 163, 211, 230
Cost effectiveness ... xi, xiii,
  15, 29, 31, 66, 113, 116,
  142-144, 146, 150, 153,
  159, 168-169, 180, 190,
  210-211, 214, 216-220,
  224, 227, 233
Cost-effectiveness
  analysis ... 163
Cost-effectiveness
  consideration ... 166
Cost-effectiveness
  considerations ... 8,
  15-16, 113, 116, 146, 154,
  161, 223, 228
Counter indication ... 141
Course ... 32
Cracking ... 42, 64, 115-116,
  149
Crisis ... 106, 108, 172, 180
Critical ability ... 51, 143
Critical attitude ... 232

Critical discussion ... 62, 196
Critical examination ... x-xi,
  xiii, 28, 41, 56, 98, 104,
  125-126, 129, 149, 152,
  158, 226
Critical study ... 93
Critical thinking ... x, 49
Criticism ... xvi, 12, 19, 45,
  47, 62, 109-110, 116, 122,
  131
Crucial experiment ... 119, 122
Crucial tests ... 145, 220
Curiosity ... 86-89, 131, 226
Custom ... xiv, 38, 68-69, 116,
  121, 123, 125, 146, 178
Customer ... 67-68
Cybernetics ... xiv, 97, 109

Data ... 37, 42-43, 46-51, 54,
  58, 89, 93, 118-119,
  145-146, 159, 169, 212,
  214, 217, 229
Data banks ... 30, 47, 58, 210,
  223
Data retrieval ... 146, 223
Deceit ... 71
Decision ... xiii, 13, 27, 29-30,
  32, 38-39, 42, 52-53, 64,
  79, 82, 84, 87-88, 94,
  100, 103, 120, 124,
  135-136, 139, 147, 157,
  181, 219, 221, 230
Decision maker ... 75, 102-104,
  106-108
Decision making ... xv, 63,
  76-78, 168, 222-223, 231
Decision process ... xiii, 153
Decision system ... 211
Decision theory ... xi, xiii, 42,
  66, 166, 169
Decision tree ... 204, 219-221

Democracy ... 70-71, 73-74, 88, 106-107, 111, 196, 232

Democratic control ... 58, 88, 106, 156, 161

Democratic process ... 60, 82, 84, 109, 176

Democratization ... 107, 233

Deviant ... 88, 198, 205

Diagnostic encounter ... ix, xii, 9-10, 18, 33, 39, 41, 57, 60-62, 65, 68, 93, 117, 142-143, 172

Diagnostic judgment ... 37

Diagnostic performance ... 38

Diagnostic practice ... 5-6, 26, 29, 40, 53, 62, 82, 125-126, 142, 145, 159

Diagnostic process ... ix, xii, 3-6, 9, 13, 25, 32-33, 39, 51, 54, 59-62, 65, 75, 78, 81, 84, 138-139, 150, 152-153, 188, 218, 225

Diagnostic service ... xv, 1-2, 14-15, 33, 39, 59, 61, 63, 65, 72, 85, 111, 113, 133, 145-146, 148, 150, 161-164, 170, 191, 209, 211, 214, 216-218, 220, 227-232

Diagnostic technology ... 11, 157

Diagnostic theory ... 17, 117, 140, 232

Diagnostics ... ix-xii, 2, 5-6, 9-12, 14-15, 17-19, 28, 34, 44, 61, 76, 113, 117, 131, 146, 156, 161, 193, 219, 224-226, 229-233

Dictionary ... 6, 31, 93

Diet ... 128, 206, 226

Differential diagnosis ... 11, 35, 132, 145-146, 152, 154, 158, 162, 187, 220, 233

Disease ... 32-35, 43, 65, 93, 101, 119, 129, 131, 133, 138, 153, 210, 215-216, 218, 221, 225, 233

Disease entities ... 43, 138, 150-151, 158, 196, 199

Disorder ... 197-199

Dissident ... 197-199

Distribution ... 46-47, 158

Diversity ... 196

Dogmatism ... 2, 78, 96, 139

Druggist ... 177

DSM ... 48, 197-208

Dual disease ... 158-159, 215-216

Dysfunction ... 204, 208

Ecology ... 99, 101

Economics ... xiii, 4, 16-17, 27-33, 39, 86-87, 89, 129, 164, 176, 210, 231

Education ... 4, 15, 17, 51, 57, 60, 80, 176, 181, 192, 230-233

Efficiency ... 12, 17, 64, 75, 90, 104, 110, 116, 118, 124-125, 132-133, 142, 145, 149, 218, 223-224, 227

Elimination ... 134, 212-213

Eliminative induction ... 8, 55

Emergency ... 70, 74, 123-124, 143, 162

Emergency-room ... 65

Empiricism ... xi, 1, 3, 13, 43-46, 85, 94, 131, 221

Enhancement ... 54, 230

Epidemic ... 82, 93-94, 133, 160-161, 168

Epidemiology ... 15-16, 65, 82-83, 92-95, 117, 130, 132-133, 144, 153, 225

Equilibrium ... 209, 229

Erewhon ... 225

Error ... 37-38, 45, 52-53, 57, 59, 72, 93, 99, 101-102, 123, 125-127, 133-136, 145, 151-152, 156, 163, 167-168, 185, 213-214, 216, 220, 222

Etiological treatment ... 162

Etiology ... 3, 12, 32, 43-46, 119, 122, 196, 215

Etiology-orientation ... 165

Evaluation ... 17, 102, 129, 166, 222

Excess ... 17, 41-42, 46, 73, 78, 89, 124, 143, 168, 172, 188, 215, 221, 230, 233

Expediency ... 70

Experiment ... 12, 25, 34, 36, 46, 87, 119, 122-123, 131, 233

Expert systems ... xii, 4-5, 11, 49, 52, 55-56, 61, 64-65, 92-93, 142-143, 148, 150-152, 159-161, 217-218, 223, 226-227

Expertise ... 9, 30, 55-56, 80

Experts ... xv-xvi, 25, 30-31, 37, 44, 54, 56, 64, 69, 79-80, 88, 135, 157, 159-160, 223, 225, 227

Explanation ... 61, 92, 96, 98, 118, 121-122, 124, 196

Explanatory power ... 118

Extreme empiricism ... xi, 43, 45-46, 52, 57, 104, 194

F.U.O. ... 8, 119

Fact ... x-xi, 1-3, 35, 43, 46-47, 52, 118, 131, 145, 148, 178, 194, 196-197, 200

Fashion ... 88, 128, 142, 226-227

Fatigue ... 124, 136

Filter ... 151, 153, 214

Flexibility ... 88-89, 146, 153, 213, 222, 229

Follow-up ... 45, 81, 129, 140, 147

Food and Drug Administration ... 51, 122

Forensic medicine ... 75

Formalism ... 12, 25, 34, 39-40, 65, 100-101, 116-117, 119, 149-150, 184, 186, 209, 213, 218-220, 224, 227, 229

Fragmentation ... 4-5, 178, 211

Framework ... ix, xiii, xvi, 12, 19, 40, 53, 67, 71, 87, 91, 118-119, 129, 140, 206, 211, 231

Frequency ... 143, 148

Fully computerized diagnosis ... 6, 29, 59-60, 64, 214, 216, 227

Function ... xi, 95-97, 102, 104-106, 119, 219

Functional disease or disorder ... 195

Game ... 86-87, 116

General check-up ... 163-164, 169, 172-173

General systems theory ... 97

Good Samaritan ... 75

Guarantee ... 19, 30, 43, 45, 55, 84, 90, 224, 232

Guidelines ... xvi, 13, 15, 18, 48, 61-62, 66, 68, 108, 131, 139, 142, 144, 163, 178, 217, 222
Guild ... 176-177
Guinea pig ... 85, 90, 131-132

Habitat ... 98
Hardware ... xvi, 156, 177, 217, 233
Heuristic ... 35, 48, 92, 95, 97, 104, 187
Hippocratic code ... 74-75, 78
Hippocratic oath ... 75, 175
Holism ... xiv, 28, 91, 95-96, 105
Homosexuality ... 74, 199, 201-202
Hospital ... 16-17, 68, 85, 109, 117, 127, 167, 171-173, 210-211, 231
Hospitalization ... 14, 33, 63, 172, 210
Hypotheses ... 1-3, 7, 9, 13, 45-51, 56, 58, 104, 118-119, 131, 139, 145, 148-149, 152, 187, 208, 223
Hypothetico-deductivism ... xi, 1, 44-50, 53, 56, 58, 200

Iatrogenic disease ... 141-143, 172
ICD ... 93, 195-197, 199
Idealization ... 1, 19, 26-28, 30-31, 33, 76, 78-79, 98-99, 116, 120, 125, 228
Image ... 153
Incentive ... 16-17, 56, 72, 130-131, 168, 212, 226, 229
Incompleteness ... 55, 95

Individualist ethics ... xiii, 68-76, 78, 82, 168, 227
Induction ... 43, 46, 55, 104, 116, 118, 148, 178
Inductivism ... 2-3, 42-49, 51, 53, 56, 58, 90, 104, 146, 148, 200, 207
Information ... 8, 30-31, 35-36, 38, 42-43, 46-47, 50-51, 60, 63, 65-66, 79-81, 87, 127, 145, 152, 156, 162, 170, 173, 193, 196, 209-212, 214-216, 219, 229, 232
Information processing ... 220
Information theory ... xiii, 152, 156, 210, 212, 214-216
Informed consent ... 9, 13, 16, 65, 75, 77-81, 83, 120, 132, 166, 231
Inner conflict ... 70, 106-107
Inoculation ... 82
Inquisition ... 79
Institution ... 51, 53, 56-57, 65, 68-69, 71-73, 82, 84, 87-88, 93-94, 121-122, 124, 134, 230, 232
Instruction ... ix, 7, 51, 86, 149, 159, 219, 233
Instrument ... 52, 63, 224
Insurance ... 14, 43, 56, 71, 129-130, 172, 231-233
Integration ... ix, xiv-xv, 1, 4-5, 15, 56, 66, 89, 129, 147, 165, 168, 211, 225, 229, 233
Interaction ... xv, 4, 14, 28-29, 43, 54, 64, 70, 84, 88, 92, 96, 98-99, 152, 217, 222, 224, 227-230, 232
Interface ... 59, 155

Internal medicine ... 177
Intuition ... ix-x, xii, 9, 17, 31, 35-38, 43, 54, 65, 143, 146-147, 151-152, 158-159, 180, 184, 188-189, 211, 214, 219, 222
Irrational practice ... 123-124
Irresponsibility ... xvi, 17, 52, 72, 121, 123-125, 127, 130, 134, 141, 228
Irresponsible error ... 45, 59
Isolation ... 1, 4, 13-15, 18-19, 28, 78, 87, 91-92, 95, 99, 102, 150

Judgment ... xiv, 36-37, 39, 63, 68, 101, 108, 147, 156-158, 212, 222
Justification ... 57, 107, 121-125, 128, 143, 223

Knowledge explosion ... 5

Language ... 184, 186, 219
Law ... 48-51, 55, 61, 68-71, 73-74, 81, 85, 116, 121-126, 129, 177, 185, 206
Law court ... 48, 51, 77, 83, 126
Leader ... 88, 127, 178
Legal framework ... 71
Legal reform ... 50, 75, 79
Legitimacy ... 17, 70, 95, 97, 230
Liberalism ... xvi, 70, 75, 79, 112
License ... 26, 70, 122, 145, 175-177, 179, 181
Lifestyle ... 15-16, 60, 205-207, 225-226, 228, 233

Limited access protocol ... 66

Malingering ... 145, 188-190, 192-194, 216
Management ... 107, 109, 133-134, 207
Market ... 41, 57, 86-87, 89, 112, 144, 185, 210, 217, 223, 229
Materialism ... x-xi, 96
Mechanism ... x, xiv, 28, 91-92, 94-98, 105, 129, 161, 190, 200
Medical administration ... xv, 18, 27, 225, 230-231
Medical association ... 27, 68, 126
Medical care ... 15, 31, 129, 161
Medical community ... 52
Medical education ... 42
Medical ethics ... xiii, xvi, 231
Medical literature ... 224
Medical politics ... 27
Medical practice ... x, xiv-xv, 5, 9, 17, 43, 70, 73, 121-122, 124, 126, 133-134, 176, 178, 205
Medical practitioner ... xvi, 7, 41, 58, 129-130
Medical profession ... xiii, 30, 53, 56, 60, 66, 127, 230
Medical record ... 36, 60, 66, 94, 134, 137, 145, 164, 175-176, 216, 231
Medical research ... 11, 85, 225
Medical school ... ix, 13, 51, 57, 61, 85, 126, 165, 219
Medical student ... xvi
Medical system ... xv, 56

Medical textbook ... 6, 9, 34, 61, 121-123, 125-126, 166-167

Medical tradition ... x, 18, 32-33, 42, 56, 60, 74, 121, 123-124, 127, 145, 178, 201

Medication ... 145, 176-177, 216

Mental disorder ... 198, 201

Mental illness ... 195

Meta-text ... 27-28, 35, 101-102

Metaphysics ... 28, 91-92, 96, 119

Method ... xiv, 8, 10, 43, 46-47, 58, 96, 145, 149, 164, 210, 214, 217-225, 227

Methodology ... 2, 53, 86, 118-119

Minors ... 177

Miracle worker ... 179-180

Modification ... 26, 32, 232

Modified individualist ethics ... 68

Monitor ... 10, 33-34, 39, 45, 52-53, 59-60, 65-66, 84, 88, 90, 109-110, 116, 134-136, 144, 156-157, 159-161, 163, 165-168, 174, 216, 218, 230

Monopoly ... 231

Moral code ... 30

Moral constraint ... 29

Moral force ... 71-72

Moral principle ... 69, 78

Moral problem ... xiii, xvi, 30, 71

Moral sentiment ... 62

Morality ... 70, 74, 78, 231, 233

Mysticism ... 28, 91

Myth ... 10, 52, 81-83

Myth of induction ... x, 2, 10, 42-43, 46, 48, 52, 55, 57, 90, 146, 173, 178

National budget ... 14, 16

National health ... 67, 142

Neglect ... ix-x, xiv-xv, 9, 11, 38, 61, 69, 82-84, 110, 166, 168, 174

Network ... 12, 64-66, 99, 227

Neurology ... 177

Neuroses ... 195

Noise ... 152-153, 158, 187-188, 191-193, 212-215, 218, 224

Noise removal ... 157-158, 188, 214

Norm ... 26, 51, 109, 112, 119, 135, 208

Normal science ... 88, 90

Nosology ... ix-x, 3, 43, 92-93, 95, 101, 196, 224

Null case ... 116-117, 139-143, 145-148

Null disease ... 142

Null hypothesis ... 49, 61, 140, 191

Null prescription ... 32

Null treatment ... 33

Nuremberg trials ... 73

Nursing ... 129

Nutrition ... 206

Obligation ... 77, 90, 122

Onion model of diagnosis ... 61

Open-endedness ... xiv, 13, 18-19, 101, 109, 115, 225

Operational research ... 106

Opportunity cost ... xiii, 211

Organicism ... 105
Over-treatment ... 17-18, 141

Para-medical
    professions ... 129
Para-text ... 27-28, 34-35, 101,
    152
Participation ... 60, 65-66, 71,
    74, 76, 82, 111, 113, 168,
    231
Paternalism ... xiii, 45, 70, 75,
    79, 155, 189, 231
Pathology ... 12, 43, 82, 129,
    174-175, 205
Patient-orientation ... 75,
    207-208, 231
Patient-oriented
    diagnosis ... 131
Pattern recognition ... xi, xiii,
    151-154, 156, 187, 213,
    217, 222
Pattern-recognition
    theory ... 54
Patterns ... 65, 104, 106,
    153-154, 158, 190-191,
    212-214, 216
Performance
    measure ... 102-103,
    117-119
Petrification ... 64
Pharmaceutic industry ... 231
Pharmacotherapy ... 143
Philosophical
    framework ... xvi, 91
Philosophy ... x, xvi, 100
Physiology ... 3, 43, 105, 129
Physiotherapy ... 129
Placebo ... 179-180
Pluralism ... 211, 232
Politics ... 18, 27-28, 70-71,
    82, 87-88, 198, 225,
    231-232

Post mortem ... 83-84, 174,
    216
Precision ... 184-186
Premature formalization ... 204
Prescription ... 3, 32-33, 43,
    48, 53-54, 62, 64, 128,
    143, 145, 171, 173,
    176-177, 205, 232
Preventive diagnostic
    service ... 163
Preventive medicine ... 5, 128,
    163
Preventive treatment ... 39
Privacy ... 26, 64, 83
Probability ... xiv, 34-38, 47,
    60, 94, 133, 139, 147-148,
    150, 152, 159, 186,
    221-222
Problem-situation ... 27
Profession ... xvi, 30, 70, 125,
    160, 175, 177-179, 212,
    222, 227, 230-232
Professional association ... 13,
    18, 177
Prognosis ... 12, 130
Programmer ... 52, 62, 64, 92,
    143, 147, 216, 227, 231
Programming ... xi, 6, 34,
    146-147, 217
Progress ... 13, 17, 41, 60,
    116, 118, 130, 160, 218,
    220, 224-225, 229, 233
Psychiatric diagnosis ... 203,
    207
Psychiatry ... 48, 177, 193,
    195, 197
Psychology ... 35-36
Psychopathology ... 129, 205
Psychophysiology ... 129
Psychoses ... 190, 195
Psychotherapy ... 129, 207
Public at large ... 82, 131, 233

Public awareness ... 12, 74, 231

Public control ... x, 12, 76, 82, 84, 89, 109, 124, 134, 145, 161

Public health ... 16, 82, 93-94, 128, 141, 145, 155, 161, 164-167, 169-170, 217, 225

Public interest ... 14, 66, 68, 70-71, 109, 171, 176

Public record ... 134

Public sector ... 87

Quality control ... x, 17, 60, 81, 161, 176, 186

Quality of life ... 225

Queuing ... xi, 58, 146, 210, 212

Randomness ... 46-48, 58, 145

Rare disease ... 159-160, 164, 191, 215-216

Rare pattern ... 191, 216

Rational disagreement ... 122

Rational planning ... xv, 14, 18, 59, 78, 80, 110, 162, 168, 209, 217, 227, 229

Rational technology ... 123

Rationality ... 85, 122-123, 126, 133, 136, 211

Reasoning ... xv, 63-64, 146, 151-152, 233

Reductionism ... 73

Redundancy ... 36, 141, 146, 152, 185, 213-214

Reform ... x, xiv-xv, 12, 56, 62, 69, 71, 74, 78, 107, 116, 122, 124, 126, 136, 166, 185, 209, 231

Regulation ... 12-13, 53, 61, 76, 134

Regulative idea ... 229

Reliability ... 35, 37, 47, 123, 137, 153, 200-201, 208, 220, 222

Repeatability ... 1, 46-47, 131, 196-197, 200-201, 203, 207-208, 211

Representative sample ... 46-47

Reproducibility ... 186, 190, 201

Research ... xv-xvi, 11, 17, 37, 41, 47, 50, 53, 55, 57, 67, 84-90, 95, 98, 118, 131, 144, 150, 162, 190-191, 220, 225-226

Responsibility ... xiii-xiv, xvi, 30, 45, 52, 57-58, 60, 64, 68-74, 76, 79, 84, 113, 122-123, 125-126, 129-131, 133, 136, 147, 168, 177-178, 227, 232

Responsible error ... 58-59, 125, 133, 185

Retrievability ... 186

Reversibility ... 69

Revisability ... 55, 65, 196, 213

Rigor ... 25-26, 29-31

Rule ... xiv, 9, 25-27, 29-31, 34, 38-39, 42, 50, 62-63, 68-70, 74, 78, 87-88, 99, 103, 107-109, 115-116, 121, 123, 131, 133-134, 196, 209, 213, 219, 222, 230

Safeguard ... xvi, 14, 17, 26, 38, 64, 88, 93

Sample ... 47, 157

Sampling ... 46-48, 94

Sampling method ... 46

Science ... x-xi, xiv, 1-3, 28, 35, 43, 45, 47, 53, 57, 60, 62, 84, 86-91, 95-96, 99, 116, 118, 122-123, 127-128, 131, 133, 185, 196, 200-201, 203-204, 207, 211, 213-214, 219, 224-226, 228, 231, 233

Science-oriented diagnosis ... 132

Scientific medicine ... 127, 157, 207-208

Scientific method ... 2, 47, 58, 118, 131

Scientific technology ... xiv, xvi, 4

Second opinion ... 170-171, 173-174, 213, 215

Self-deception ... 181

Self-diagnosis ... 38, 57, 143, 181

Self-interest ... 231

Semi-autonomous teams ... 106-107

Side-effects ... xiv, 39, 141-145, 165, 180, 185

Sign ... 32, 34, 137, 215, 222

Signal ... 212-214

Simulation ... 64, 115-116, 148-151, 184, 210, 218, 222-224

Social and political philosophy ... xvi, 70

Social factors ... 13-16, 38, 51, 60, 71-72, 75-76, 97, 122, 125-126, 145, 201, 225

Social science ... ix, 134, 204, 228

Social technology ... ix

Sociotherapy ... 193

Soft and hard medicine ... 100, 108, 129, 205, 232-233

Software ... 142-143, 150, 152, 156, 161, 217, 223-224, 226, 232

Specialists ... xiv, 15, 30, 38, 55-56, 89, 159-160, 177, 221

Specialization ... x, xiv-xvi, 4-5, 15, 17, 31, 80, 87-89, 149-150, 165, 179

Standard clinical pathological conference ... 174

Standard clinical survey ... 174

Standardization ... 50-51, 60, 93, 113, 137, 150, 184, 186, 195-197, 201

Standards of health ... 225

Statistical considerations ... 35

Statistical hypotheses ... 13

Statistical method ... 46, 221

Statistical theory ... 46, 222

Statistical weight ... 218

Statistics ... 16, 35, 38, 46-47, 92-95, 123, 133, 140, 142, 145-147, 153, 161, 169, 192, 207, 211, 221, 223

Storage ... xi, 58, 207, 210, 212

Subjective probability ... 47

Subjectivism ... 47-48, 51, 94-95, 186

Support ... 3, 8, 134, 138

Surgeon ... 75, 94, 111, 177-179

Surgery ... 62-63, 125, 128-129, 134-135, 141, 178

Survival ... 72, 108

Symptom ... 3, 32, 34, 43-44, 101, 119, 137-138, 152, 195, 212, 215, 222

Symptomatic
  treatment ... 162, 165
Symptomatology ... 191
Syndrome ... 32, 34-35, 43,
  119, 138-139, 151
Systemism ... xi, 73, 75, 82,
  190, 195
Systems ... 107
Systems analysis ... xi, xiii,
  15, 59-60, 66-68, 95, 97,
  100, 102-112, 117-118,
  130, 132-133, 211,
  218-219, 224, 228
Systems approach ... xi, 27,
  92, 95-97, 102, 104,
  108-109, 115, 155, 187,
  219
Systems engineering ... 97
Systems theory ... xi, 4, 27-28,
  34, 91, 95, 97, 102, 126

Tacit knowledge ... 44
Technician ... 54, 70
Technique ... xv, 11-12, 14,
  26, 32, 34-35, 38, 106,
  134, 153, 211, 217-224,
  228, 233
Technological innovation ... 77,
  87
Tentativeness ... xiv, 13,
  27-28, 32, 51, 65, 78, 92,
  96, 99, 101-102, 108-109,
  152, 154, 216, 223
Terminal patient ... 180
Test ... 1-3, 11-13, 19, 25,
  46-51, 53, 57-58, 79-80,
  83, 93-94, 104, 121-123,
  128-130, 133-135, 138-139,
  145-146, 149, 151, 153,
  162-164, 170, 210, 213,
  217, 220-221, 230, 233
Testability ... 131

Tomography ... xi, 6, 11, 41,
  54, 63, 71, 157, 217
Tool ... xi, 54
Torture ... 18, 74
Trade secret ... 87
Tradition ... xvi, 125-126
Transaction analysis ... 193
Translation ... 184
Treatment ... xiv-xv, 3, 5,
  9-10, 12-15, 17-18, 30,
  32-35, 39, 43, 45, 48, 62,
  64-65, 71, 76-78, 81, 85,
  111, 119, 121, 123,
  127-129, 148, 154, 162,
  170, 206, 209-210, 215,
  225, 227, 232-233
Trust ... 134

Unanimity ... 196, 201, 204,
  211
Unanimity and diversity in
  science ... 196
Uncertainty ... 25, 34-35
Under-treatment ... 143
Unification ... 88
Updating ... 9, 30-31, 214
Utility ... 60, 65
Utility function ... 7
Utopia ... 61, 209, 227-228

Victimless crimes ... 74

Weight function ... 16-17

Episteme

A SERIES IN THE FOUNDATIONAL,
METHODOLOGICAL, PHILOSOPHICAL, PSYCHOLOGICAL, SOCIOLOGICAL, AND
POLITICAL ASPECTS OF THE SCIENCES, PURE AND APPLIED

1. W.E. Hartnett (ed.): *Foundations of Coding Theory.* 1974   ISBN 90-277-0536-4
2. J.M. Dunn and G. Epstein (eds.): *Modern Uses of Multiple-Valued Logic.* 1977
ISBN 90-277-0747-2
3. W.E. Hartnett (ed.): *Systems: Approaches, Theories, Applications.* 1977
ISBN 90-277-0822-3
4. W. Krajewski: *Correspondence Principle and Growth of Science.* 1977
ISBN 90-277-0770-7
5. J.L. Lopes and M. Paty (eds.): *Quantum Mechanics, a Half Century Later.* 1977
ISBN 90-277-0784-7
6. H. Margenau: *Physics and Philosophy.* Selected Essays. 1978   ISBN 90-277-0901-7
7. R. Torretti: *Philosophy of Geometry from Riemann to Poincaré.* 1978
ISBN Hb 90-277-0920-3 / Pb 90-277-1837-7
8. M. Ruse: *Sociobiology: Sense or Nonsense?* 2nd ed. 1985
ISBN Hb 90-277-1797-4 / Pb 90-277-1798-2
9. M. Bunge: *Scientific Materialism.* 1981   ISBN 90-277-1304-9
10. S. Restivo: *The Social Relations of Physics, Mysticism, and Mathematics.* Studies in Social Structure, Interests, and Ideas. 1983
ISBN Hb 90-277-1536-X / Pb (1985) 90-277-2084-3
11. J. Agassi: *Technology.* Philosophical and Social Aspects. 1985
ISBN Hb 90-277-2044-4 / Pb 90-277-2045-2
12. R. Tuomela: *Science, Action, and Reality.* 1985   ISBN 90-277-2098-3
13. N. Rescher: *Forbidden Knowledge* and Other Essays on the Philosophy of Cognition. 1987   ISBN 90-277-2410-5
14. N.J. Moutafakis: *The Logics of Preference.* A Study of Prohairetic Logics in Twentieth Century Philosophy. 1987   ISBN 90-277-2591-8
15. N. Laor and J. Agassi: *Diagnosis: Philosophical and Medical Perspectives.* 1990
ISBN 90-277-0845-X
16. F.P. Ramsey: *On Truth.* Original Manuscript Materials (1927–1929) from the Ramsey Collection at the University of Pittsburgh, edited by N. Rescher and U. Majer. 1990
ISBN 0-7923-0857-3

KLUWER ACADEMIC PUBLISHERS – DORDRECHT / BOSTON / LONDON